VEDIC COSMOGRAPHY IN A MODERN CONTEXT

VEDIC COSMOGRAPHY IN A MODERN CONTEXT

VIRODHA-PARIHĀRA REVISITED

A collaborative project of associates of the Bhaktivedanta Institute for Higher Studies and Richard L. Thompson Archives

Editors:
S. E. Kreitzer, Prishni Sutton, Akhaṇḍadhī dāsa

Contributing Authors:
Prishni Sutton, S. E. Kreitzer (S. D. Muni), Akhaṇḍadhī dāsa, Bhūrijana dāsa, Richard L. Thompson (Sadāpūta dāsa), Pracārānanda dāsa, Antardwīp dāsa, Murli Gopal dāsa, Emmanuel Namamrta Greene, Prahlad Griffith, Krishna Swarūpa dāsa

RLTA PRESS
Alachua, Florida

Cover and book design: Eight Eyes, eighteyes.com

On the Cover: "Astronomy: the open air observatory at Delhi" (1797). Colored aquatint by Thomas Daniell, 1815. This is one of the five observatories constructed by Mahārāja Sawai Jai Singh II (1688–1743) of Jaipur. Image Credit: Wellcome Collection. See: wellcomecollection.org/works/s3dpx599. The circular image framing the title on the front cover illustrates a globe commissioned by Jai Singh in the early 1700s. It includes geographic features of South Asia. For more information please see page 50 of Richard L. Thompson's *Mysteries of the Sacred Universe* (2000), and page 71 of this text. The image of Jai Singh's observatory at Benares on the back cover is also featured on Thompson's *Vedic Cosmography and Astronomy* (1989). It was originally published in *Philosophical Transactions of the Royal Society of London* 67 (1777), 608, Tab XII.

Images on pages *vi*, 13, 16, and 197–223 courtesy of The Bhaktivedanta Book Trust International, Inc. www.krishna.com. Used with permission.

The editors of *Vedic Cosmography in a Modern Context:* Virodha-parihāra *Revisited*, would like to offer their profound appreciation to the Bhaktivedanta Book Trust trustees. Their dedicated support of the activities of the Bhaktivedanta Institute for Higher Studies and Richard L. Thompson Archives facilitated the production of this volume.

For information on the Bhaktivedanta Institute:
https://bihstudies.org
info@bihstudies.org

For information on the Richard L. Thompson Archives:
https://richardlthompson.com/
info@richardlthompson.com

Second Edition
ISBN: 978-1-959829-97-3 (hardcover)
ISBN: 978-1-959829-04-1 (softcover)

Dedicated to A. C. Bhaktivedanta Swami Prabhupāda, whose translations and commentaries on *Śrīmad-Bhāgavatam* and other Vaishnava literature have been popularized by his followers around the world. This book is also dedicated to Prabhupāda's spiritual master, Bhaktisiddhānta Sarasvatī Ṭhākura, whose Bengali translation and commentaries served as the foundation for Prabhupāda's work on the Fifth Canto. Bhaktisiddhānta Sarasvatī has also been noted as one of India's leading *jyotiṣa brāhmaṇa paṇḍitas* in the late 19th and early 20th centuries.

Even those who are constantly engaged in sinful acts are all corrigible to the standard of perfect human beings if they take shelter of the devotees of the Lord. Jesus Christ and Muhammad, two powerful devotees of the Lord, have done tremendous service on behalf of the Lord on the surface of the globe.

A. C. Bhaktivedanta Swami Prabhupāda
Śrīmad-Bhāgavatam 2.4.18, Purport

His Divine Grace A. C. Bhaktivedanta Swami Prabhupāda, Founder-*Ācārya* of the International Society for Krishna Consciousness, speaking with a disciple in Geneva, Switzerland in 1974, with a globe on the desk in his quarters.

Contents

SECTION III
Contemporary Discussions: East and West

SECTION IV
Conclusion and Appendices

Foreword

It is often, if not generally, the case that questions which initially seem to be simple and straightforward turn out to be cause for difficult discourse leading to much consternation as conclusive truth is sought. "Is the earth flat, or is it spherical?" is such a question. Probably most of us who are now reading this book grew up with the understanding that the earth is spherical – like all the other planets in the "Solar System." I recall the ditty we learned in elementary school: "In fourteen hundred ninety-two / Columbus sailed the ocean blue ... Columbus knew the world was round / So he looked for the East while westward bound ..." And so, one thought, the matter was settled, thanks in large part to Christopher Columbus's brave resolve, even if he reached less than half way around the globe to the India he had hoped to reach.

But then, years later, some of us were introduced to a radically different worldview, embodied in the corpus of ancient Indian texts broadly referred to as "Vedic," including the *Purāṇas,* a rich genre of works in classical Sanskrit that contain brilliant, subtly challenging philosophy alongside elaborate and seemingly fantastical cosmography. One might have been tempted to dismiss the cosmography as, well, "outlandish" (as did – famously or infamously – the early nineteenth-century British politician Thomas Macaulay). But for one who would seriously seek to comprehend the truth of Vedic philosophy, under the guidance of authentic teachers in disciplic succession, it would not be an option to reject any portion of the Vedic corpus on the basis of initial incomprehension. Further, these descriptions – of concentric ring-like "islands" separated by "oceans" of liquor, clarified butter, (never souring) milk, and yogurt, and of "mountains" from which flow rivers in four directions, the muddy banks of which, when mixed with the pulp of (elephant size) *jambū* fruits create gold – were taken seriously by early Indian scholars who were also fully engaged with modern-like cosmological observation of spherical planets, including Earth, traversing space in mathematically predictable fashion (mainly for calendrical and astrological purposes). That such scholars did not reject such accounts is interesting, potentially illuminating, but possibly consternating.

The authors of the present volume – each highly qualified in their respective fields of scholarship, including Purāṇic esoteric cosmography and exoteric observable astronomy – seek to address this intriguing locus of inquiry. The aim is not to claim some final, definitive resolution or harmonization of the two apparently conflicting cosmographies or worldviews. Rather, the aim is to bring together and organize the varied traditional (Vedic-Purāṇic) sources with their traditional Sanskrit commentarial exegeses along with the similarly varied responses to those sources by similarly traditional *jyotiṣa* (astrological) pandits as well as several contemporary, carefully worked out responses. What emerges from this assembly of resources is a highly valuable knowledge base for anyone who intends to seriously delve further into the subject. Also, it serves those of us who seek an overview of the present state of the discourse with sufficient information – both wide-angle and close-up – to be more than adequately informed on the subject to be able to discuss it intelligently with a variety of audiences.

I dare say that some readers of this work – those who think that a certain present-day popular "flat earth" geography supports Purāṇic geography and cosmography – may be disappointed. As the authors show, a careful examination of all the sources accepted as Vedic *śabda-pramāṇa* evidence do not discount the sphericity of the earth. Rather, they show that one must take into account the entire gamut of scriptural evidence along with the authorized commentarial tradition, as well as the further statements of saintly persons (*sādhus*) from within the cultural tradition. In other words, one may well find support for flatness (as Bhū-maṇḍala is described), and, one may – following the tradition – find it necessary and proper to acknowledge the earth (like other heavenly bodies) to be a globe.

Such a both-and approach is at the heart of this presentation. Thus, one may reasonably be compelled to acknowledge the comprehensiveness that our authors and editors have provided on this subject. This approach follows the Vedic – in particular the Vedāntic – practice of hermeneutics (the disciplined and appropriate interpretation of scripture), whereby one must take all relevant scriptural statements into account, and one must also give due attention to observation and reason. The result of this sort of careful analysis may well be unexpected: we come to the subject with our preconceptions, and then we may be surprised by the conclusion – or possibly lack of conclusion – that we come to.

By this kind of interpretive exercise, we may feel moved to reflect on the broader hermeneutical issues – questions about the process and goals of interpretation – that this inquiry raises. Fortunately, it is possible to undertake such reflection systematically. This has been shown by the authors of a recently developed hermeneutic system and course for learning to apply this system for members of the International Society for Krishna Consciousness (ISKCON).[1] The serious practitioner is therein urged to embark on a "path" of analysis that begins with self-reflection about one's own qualities and qualifications for pursuing the difficult work of authentic interpretation at hand. This is not the place to elaborate on this process. Suffice it to say, the very first of six requisite qualities we are asked to be equipped with (before taking further essential steps in the process) is "humility and service mood." And so, I'm impelled to ask myself, as I read through this important presentation, which provides luminous examples of hermeneutical practice, whether my own stock of humility and service mood pass muster at the starting point for grasping the issues at hand. What I find is that this study of apparently differing traditional and contemporary cosmographies, with all the associated ancient and modern interpretations, challenges me to make the question of humility and service attitude an ongoing one. The vastness of just this one planet on which I find myself defies my comprehension, however much I may attempt to apply my discernment to the task.

And so, I'm inclined to take refuge in the acknowledgment, which our authors call attention to in their Conclusion, that even Lord Brahmā "lacks perfect knowledge about this topic." This awareness does not mean, however, that we should abandon our endeavors to understand, recognizing all the while that our purpose is, initially and finally, to serve truth – or indeed Truth – both of which are surely better served by the ongoing pursuit of understanding.

—Krishna Kshetra Swami
Dean of Studies at Bhaktivedanta College (Septon, Belgium)
Research Fellow of the Oxford Centre for Hindu Studies (OCHS)

Acknowledgments

This book has come about as a result of the valuable work of several important contributors. First and foremost, a special thanks needs to be given to Bob Cohen, the Director for the Bhaktivedanta Institute for Higher Studies (BIHS). His tireless work encouraging and supporting countless devotees, and in particular for fostering projects such as these, which aspire to engage disciplined scholarship with the Gaudiya Vaishnava tradition on topics pertaining to religion and science. Akhaṇḍadhī dāsa, another board member of the BIHS as well as a member of the Exhibit Committee for the Temple of the Vedic Planetarium (TOVP), initially conceived of a project like this to help explore the richness of the Purāṇic tradition in light of concerns engaging the TOVP. He also contributed to writing portions of the draft manuscript, in particular to the introduction and conclusion. Prishni Sutton developed the initial outline and conceptual draft, while taking responsibility for various production concerns. She further developed content for a number of the chapters, in particular Chapter 9 "Models of Bhū-maṇḍala within ISKCON Discourse." S. E. Kreitzer (SD Muni) worked on the final manuscript and in particular wrote Section II on the *jyotiṣa* tradition, as well as contributing essays to the Appendix treatment on Śrī Vaṁśīdhara's *Bhāgavatam* commentary.

The work of Richard L. Thompson (Sadāpūta) proved a source of inspiration as an example of engaging a thoughtful approach coupled with rigorous research and personal experience in the professional sciences. His essay, "““Bhū-maṇḍala, or Middle Earth" from his pioneering work, *Vedic Cosmography and Astronomy* (1989), is offered in the Appendices. Likewise many thanks to Bhūrijana dāsa, whose essay "A Lightning-Quick Tour of Canto Five" presents an insightful philosophical overview combined with analysis of śāstric references relevant to contemporary Purāṇic discourse.

Other important contributors include Jonathan Banks (Pracārānanda dāsa) and Mauricio Garrido (Murli Gopal dāsa) for their invaluable work developing content for Chapter 10, to which Adam Carroll (Antardwīp) also offered feedback along with his essay, "Evidence for a Spherical Earth." Emmanual Namamrta Greene and Prahlad Griffith each offered essays about personal experiences that helped to bring contemporary concerns "down to earth." Special thanks to Krishna Swarūpa for his Sanskrit translations used in "Introduction to Śrī Vaṁśīdhara and His Sanskrit Commentary."

Gratitude needs to be expressed to the many colleagues, friends, and members of the devotional community who generously offered time to critique various drafts of the manuscript. An enthusiastic "thank you" goes out to the following (in alphabetical order): Devamrita Swami, Doug Watson, Drutakarma (Michael Cremo), Hridayananda das Goswami (Howard Resnick), Jayadvaita Swami, Jonathan Edelmann, Krishna Abhishek Ghosh, Krishna Dharma, Krishna Kripa, Krishna Kshetra Swami (Kenneth Valpey), Krsna Bhajana, Oliver Zambon, Pavaneshwar, Prem Gauranga, Radha Mohan, Rajasekhara, Satsvarupa dasa Goswami, Sulabha, Tamraparni, Vaishnav, Vrindavana Priya, and Yogesvara (Joshua Greene). Sincere apologies to any oversight of important contributions that have been overlooked.

All quotations from the *Bhagavad-gītā* (BG), *Śrīmad-Bhāgavatam* (SB), and *Caitanya-caritāmṛta* (CC) used in this project, are referenced to editions of translations and commentaries by A. C. Bhaktivedanta Swami Prabhupāda and published the Bhaktivedanta Book Trust. Artwork and other items courtesy of the Bhaktivedanta Book Trust International, Inc. (www.Krishna.com), are used with permission.

Introduction

The Fifth Canto of the *Śrīmad-Bhāgavatam* (or *Bhāgavata Purāṇa*) describes an enormous structure called Bhū-maṇḍala, or "circle of the earth," which depicts a nearly 4-billion-mile wide cosmic plane comprised of concentric rings, that bifurcates the universe.[1] At the center of this *maṇḍala* is the island of Jambūdvīpa with a diameter of 800,000 miles (~1.3 million km). Its southern region, Bhārata-varṣa, is identified as a place of habitation, or *loka*, where human beings engage in fruitive activity. Thus, in Purāṇic literature, Bhārata-varṣa has been associated with the earth on which we live.

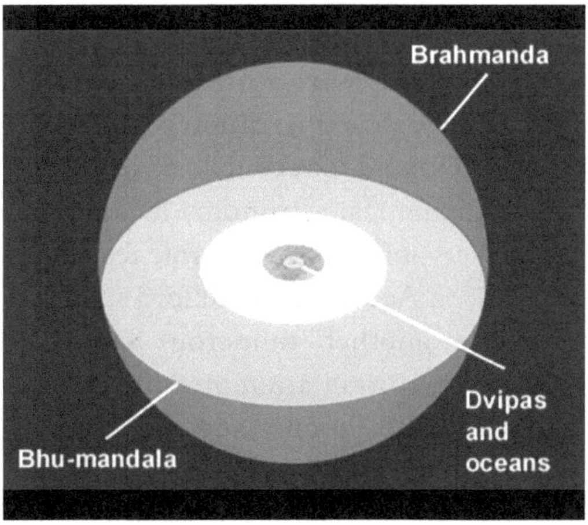

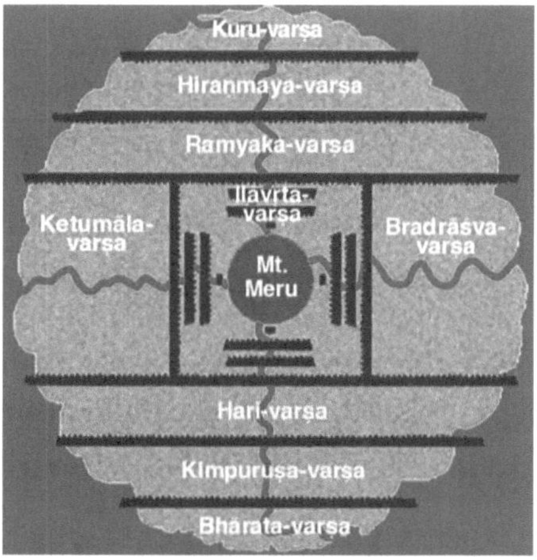

Figure 1 (left): Model of the Brahmāṇḍa showing the universal shell, Bhū-maṇḍala, and Antarikṣa. **Figure 2** (right): Map of Jambūdvīpa, with mountain ranges shown in black, the base of Mount Meru shown in the center, and Bhārata-varṣa as the southernmost region.

Without doubt, descriptions such as these offered in the *Bhāgavata Purāṇa* appear exceedingly different from anything taught in elementary school geography. Of note, neither do they correspond well with the ancient traditions of observable astronomy identified with either East or West. For example, on the Indian subcontinent, *jyotiṣa brāhmaṇas*[2] intimately familiar with Vedic cultural traditions also worked with Sanskrit mathematics and astronomy. They were keenly aware of the apparent dichotomy between the esoteric cosmology offered in the *Purāṇas* and the exoteric visible universe whose accurate charting proved essential for numerous religious activities, all of which were also described in Vedic accounts. Many of these *jyotiṣa brāhmaṇas* have also been recognized as leading pioneers in the global development of astronomy and mathematics.

This work will seek to clarify a variety of questions involving the apparent conflict between empirical astronomy and Purāṇic cosmography, while utilizing perceptions of the earth as an example. Of note, such concerns are hardly unique to contemporary readers of Purāṇic literature aware of modern astronomical achievements.[3] In other words, this is not a question of a sacred tradition in conflict with modern science.[4] After all, the basics of visible astronomy have remained constant throughout the ages. Even advocates for heliocentric versus geocentric models agree upon the same observable geometric motion of the heavens when viewed from the surface of the earth.[5]

The subtitle of this work, *Vedic Cosmography in a Modern Context:* Virodha-parihāra *Revisited,* is a nod to a cultural tradition with roots drawing from the ancient world that attempted to reconcile observable accounts identified with Sanskrit mathematical astronomy located in *jyotiṣa* texts, with the apparently unobservable esoteric cosmic features described in literature such as the *Purāṇas*. According to Oxford scholar Christopher Minkowski, "in one way or another" numerous Sanskrit commentators who identified with this tradition attempted "to demonstrate the non-contradiction between the Purāṇic and Siddhāntic models."[6]

While historians acknowledge that these apparent contradictions "were already known by the authors of the earliest *Siddhāntas*,"[7] they credit Lallācārya, author of the *Śiṣya-dhī-vṛddhida-tantra*, or "Treatise for increasing the intelligence of students," and whose work may date as early as 748 CE,[8] for establishing a "'standard' accommodation" model attempting to reconcile the two that was influential throughout the Middle Ages as adopted by Bhāskarācārya and others. Lallā's model featured a number of Purāṇic descriptions integrated within a representation of the earth globe that leaned upon the influence of the *jyotiṣa* tradition. Then in the Early Modern period, in a noticeable shift that began around the time of Śrī Caitanya Mahāprabhu, leading *jyotiṣa* astronomers such as Jñānarāja (fl. 1503) and his sons began to focus their analyses "primarily as defenders of Purāṇic authority and validity."[9] Minkowski notes,

> One finds in the astronomers of this period evidence of a growing interest in integrating their scientific method with the fully elaborated pramāṇa system of proof developed in the principal śāstras, especially in logic. Here it is the status of textual authority (śabdapramāṇa) in particular that is a matter of urgency—the canonical status of the Purāṇas and Siddhāntas, their hierarchical relation, and the degree of autonomy accorded to the other pramāṇas, that is, logic (anumāna) and experience (pratyakṣa).[10]

When considering a term like *virodha-parihāra* with reference to such activities,[11] the Monier-Williams Sanskrit-English Dictionary offers the formal definition: "removal of incompatibility, reconciliation." Jñānarāja's son, Sūryadāsa (1507–1588),[12] used *virodha-parihāra* in the name of a concluding chapter for his astronomical work, *Siddhāntasaṃhitāsārasamuccaya* (1583).[13] This chapter directly addressed "removing the contradiction between Jyotiḥśāstra and the Purāṇas," using as its Sanskrit title, *jyotiḥśāstrapurana virodhaparihāra.*[14] Another scholar, Kevalarāma Jyotiṣarāya (d. 1782),[15] noted as a the "chief astronomer"[16] in the court of "the astronomer prince of India" Mahārāja Jai Singh (1688–1743),[17] also used this term in a work commissioned by Singh with the Sanskrit title, *Bhāgavatajyotiṣayor Bhūgola virodhaparihāra.* This translates to English as, *Removal of the Disagreement between the Bhāgavatpurāna and Astronomy Concerning the Sphere of the Earth.*[18] (Another treatise with a similar title, *Bhūgolakhagola virodhaparihāra,* written a few decades later by Nandarāma Miśra "expanded and revised" Kevalarāma's work.)[19] Kevalarāma proposed in his *virodha-parihāra* treatise "to reconcile the two models," one involving an "earth" associated with the massive Purāṇic Bhū-maṇḍala plane, and the other model involving the more familiar Siddhāntic concept of "a small spherical earth," albeit in Kevalarāma's presentation the spherical earth is "floating above the Purāṇic one." Kevalarāma's work mathematically considered the motion of planets from a geocentric perspective, with the planetary orbits "pass[ing] through the surface of the flat Purāṇic earth [the Bhū-maṇḍala plane] as they revolve."[20] This could suggest that on some level, Kevalarāma's two models share perceivable space. As for this present project, *Vedic Cosmography in a Modern Context:* Virodha-parihāra *Revisited*, it aims to consider a well-established discourse drawn from within this long tradition along with contemporary discourse acquainted with more recent evidence drawn from the natural world, while utilizing the observable earth as a heuristic pivot.

Sometimes readers of the *Purāṇas* who accept its authenticity and feel intrigued by descriptions such as that of a massive 4-billion-mile wide Bhū-maṇḍala plane, subsequently explore arguments discounting the spherical nature of the earth. In a response to such concerns, this presentation focuses primarily upon evidence from Vedic sources accepted as *śabda-pramāṇa* – guru, *sādhu*, and *śāstra* – with additional references drawn from the cultural legacy that upholds these conventions. Hence, this work will focus upon the teachings of A. C. Bhaktivedanta Swami Prabhupāda as the Founder-*Ācārya* of ISKCON (guru), the saintly personalities recognized within the cultural tradition (*sādhu*), and the scriptural literature authorized by them (*śāstra*) in an examination of the contrast between the cosmic portrayals described in the *Purāṇas* and the astronomy experienced via standard human perception over the millennia. This work examines Vaishnava teachings (*pramāṇa*) that consider the earthly realm in two fundamentally different ways – as the *gola* (or globe) model, and also as a realm connected with the vast Bhū-maṇḍala plane, identified with Bhārata-varṣa situated on the island of Jambūdvīpa within Bhū-maṇḍala's central region.

As noted, awareness of the conflicting perspectives identified with Purāṇic cosmology and *jyotiṣa* mathematical astronomy is not something unique to the modern era. After all, *śāstra* describes numerous features attributed to subtle realms of higher-dimensional phenomena in apparent contrast to standard perception. These phenomena involve all varieties of experience, both within an otherwise observable universe and that which may exist beyond it. The tradition readily acknowledges that much of this would prove inaccessible to most of human society, limited as they may be by relative conditioning. In other words, while

Purāṇic commentary upholds the possibility of a direct experience of extraordinary phenomena, that would be also something considered as likely having a low probability for those enamored within a relative state, in which their original unhindered nature has been covered by layers of material conditioning.

Along similar lines, there is the traditional parable identified with the Indian subcontinent of the blind men "observing" an elephant that can also suggest the limitations of human conditioning. In this allegory, each natural philosopher becomes convinced that their own personal experience, however nominal, surely encompasses the whole phenomenon. The general idea is that when one blind man examines the elephant's leg, he imagines a tree trunk. When another blind man considers its tail, he imagines a rope, and so on. How can one animal possess such conflicting qualities? With firm conviction, each of the blind men enthusiastically contradicts the testimony of the others.[21] While recognizing such a potential state of affairs, followers of the Vedic path readily appreciate the relevance of acknowledged sources that authoritatively consider a comprehensive perspective drawn from the tradition *(aitihya)*. That, coupled with heartfelt inquiry, are considered key to realizing principles embedded in śāstric accounts.

Both the mathematically quantifiable observable astronomy found in *jyotiṣa* texts such as the *Sūrya-siddhānta* and the seemingly imperceptible cosmological accounts located in Purāṇic texts exemplified by the *Śrīmad-Bhāgavatam* are considered authoritative *pramāṇa* within a Vedic cultural context. Both need to be acknowledged. Yet while comparing the two, Śrīla Jīva Gosvāmī describes in his *Tattva-sandarbha* that Vaishnavas consider śāstric texts focused on *bhakti* to be the more authoritative. Having said that, the Vaishnava *ācāryas* also recognize the *Sūrya-siddhānta*.

As a notable example, Śrīla Prabhupāda's spiritual master, Śrīla Bhaktisiddhānta Sarasvatī Gosvāmī, earned recognition as a well-established *jyotiṣa paṇḍita* for which he received the title "Siddhānta Sarasvatī." He was especially honored for his Bengali translation and commentaries on the *Sūrya-siddhānta* and *Siddhānta-śiromaṇi*, two preeminent Sanskrit works that mathematically describe the earth as a globe floating in space. Of significance, Śrīla Prabhupāda wrote with pride about these achievements of his spiritual master.[22] Meanwhile, with regard to preserving the Purāṇic tradition, Prabhupāda drew nearly exclusively upon Bhaktisiddhānta Sarasvatī Gosvāmī's Bengali translation and commentaries for his own work presenting the Fifth Canto of the *Śrīmad-Bhāgavatam* in English to an international audience.

Without question, the *Śrīmad-Bhāgavatam,* known as the "literary incarnation of God," portrays a profoundly different description of the universe than the *jyotiṣa* tradition of observable astronomy. Its purpose is also significantly different. The *Bhāgavatam* presents the cosmos as the *virāṭ-rūpa*, a universal form intended for meditation as an initial step in God realization.[23] In fact, that is precisely how Mahārāja Parikṣit framed his inquiry when he requested Śukadeva Goswāmī to speak on Purāṇic cosmology. According to Maharaja Pariksit:

> When the mind is fixed upon the Supreme Personality of Godhead in His external feature made of the material modes of nature – the gross universal form – it is brought to the platform of pure goodness. In that transcendental position, one can understand the Supreme Personality of Godhead, Vāsudeva, who in His subtler form is self-effulgent and beyond the modes of nature. O my lord, please describe vividly how that form, which covers the entire universe, is perceived.[24]

In this verse, Mahārāja Parikṣit clearly articulates why the *Bhāgavata Purāṇa* is given priority within the Gaudiya Vaishnava tradition. It is a *śāstra* that both Śrīla Prabhupāda (SB 1.2.4, Purport) and Śrī Sūta Gosvāmī (SB 12.13.18) have glorified as "the spotless *Purāṇa*" for reasons similar to those of Mahārāja Parikṣit when he described how the *Bhāgavata Purāṇa* brings a practitioner "to the platform of pure goodness" in a manner that facilitates attaining "that transcendental position" where "one can understand the Supreme Personality of Godhead."[25]

Śrī Caitanya Mahāprabhu offers a similar assessment of the *Bhāgavatam* in *Śrī Caitanya-caritāmṛta, Madhya-līlā* 25.143–45:

> The meaning of the *Vedānta-sūtra* is present in *Śrīmad-Bhāgavatam*. The full purport of the *Mahābhārata* is also there. The commentary of the *Brahma-gāyatrī* is also there and fully expanded with all Vedic knowledge.

> *Śrīmad-Bhāgavatam* is the supreme *Purāṇa*, and it was compiled by the Supreme Personality of Godhead in His incarnation as Vyāsadeva. There are twelve cantos, 335 chapters and eighteen thousand verses. The essence of all Vedic literature and all histories has been collected in this *Śrīmad-Bhāgavatam.*[26]

This present project, *Vedic Cosmography in a Modern Context*, begins with an exploration of Purāṇic cosmography based on a review of Śrīla Prabhupāda's statements referring to the Fifth Canto, as well as statements on everyday perceptions of the earth. It goes without saying that Prabhupāda held absolute confidence in the *Śrīmad-Bhāgavatam* whenever he mentioned the earth as a globe floating in an "ocean of air."[27] Dozens of excerpts taken from Prabhupāda's lectures, morning walks, and room conversations offer similar descriptions of the earth as a planetary sphere.[28] Meanwhile, locating references to the observable earth as a flat disk remains a challenge. Śrīla Prabhupāda specifically encouraged his followers holding PhD degrees to studiously examine these texts, with special interest on the esoteric nature of the Fifth Canto, which does not appear to match the way we ordinarily experience the physical world.[29] This chapter will also consider the teachings of Śrīla Prabhupāda's spiritual master, Śrīla Bhaktisiddhānta Sarasvatī Ṭhākura, and his direct involvement in preserving both the legacies of Purāṇic cosmology and Sanskrit mathematical astronomy, within a modern context.

The *ācāryas* and *sādhus* identified with the Vaishnava tradition provide valuable insight concerning the import of Purāṇic cosmology. After a brief investigation of different arguments that they utilized, Chapter 2 examines Vaṁśīdhara Paṇḍita's *Bhāgavata* commentary in his *Bhavārtha-dīpikā-prakāśa*, which Prabhupāda considered from a list of commentaries by respected Vaishnava *ācāryas*. Śrī Vaṁśīdhara's work discusses the relationship between the Bhū-maṇḍala disk described in the Fifth Canto and the earth globe more commonly identified with ordinary experience.

Chapter 3 will compare the spherical nature of the earth as considered in both the *Śrīmad-Bhāgavatam* and the *Sūrya-siddhānta*. It is also worth noting that throughout the *Bhāgavata Purāṇa* and related texts there are numerous examples of *jyotiṣa brāhmaṇas* performing functions that represent a tradition that historically quantified the spherical dimensions of the earth.

Chapter 4 through 8 will explore the long conversation that took place within the literature of the *jyotiṣa brāhmaṇas*. This canon also analyzed apparent contradictions between Sanskrit mathematical astronomy and the cosmology of the *Bhāgavata Purāṇa*. Pandits within this tradition proved well aware of precise mathematical astronomical calculations (integral to Vedic astrology as well), alongside Purāṇic accounts that can appear to transcend ordinary experience. Luminaries identified with this legacy include Āryabhaṭa, Brahmagupta, and Bhāskarācārya, all well known also within the global history of astronomy and mathematics. Meanwhile, leading *jyotiṣa brāhmaṇas* such as Jñānarāja and his sons Sūryadāsa and Cintāmaṇi, were contemporaries of Śrī Caitanya Mahāprabhu. In addition to producing authoritative works on Sanskrit mathematics and astronomy, Jñānarāja's family wrote at length on topics identified with the Vaishnava tradition. More recently, Śrīla Prabhupāda's spiritual master, Śrīla Bhaktisiddhānta Sarasvatī Ṭhākura, had been similarly appreciated as a leading *jyotiṣa* scholar prior to his taking *sannyāsa* in 1918 and establishing the Gaudiya Math. As previously mentioned concerning the Purāṇic tradition, it was Śrīla Bhaktisiddhānta Sarasvatī's commentary on the Fifth Canto that Śrīla Prabhupāda utilized nearly exclusively as the foundation for his own work on the subject.

Initiatives exploring contrast, correspondence, and potential synergy between observable evidence and Purāṇic accounts have proven to be an established part of the tradition. Chapter 9 examines several members of the ISKCON community who have been active in exploring these issues within a contemporary context. Published authors include Richard L. Thompson (Sadāpūta dāsa), Danavir Goswami, Ashish Dalela (Rsiraja Das), and Santhoju Pandu (Pavaneshwar Das), among others. Also examined will be research from the Temple of the Vedic Planetarium, the Bhaktivedanta Institute for Higher Studies, as well as independent scholars. Similarly, the Appendices offer a variety of contemporary Purāṇic commentaries that stand on their own. These examples help illustrate some of the breadth of recent work undertaken on this subject.

Though numerous aspects of the Purāṇic universe can also appear compatible with views of the earth as a globe, popular debates sometimes consider contemporary flat earth perspectives for modeling Bhū-maṇḍala. Advocates for arguments identifiable with 19th-century evangelical initiatives, for example, propose that their model offers a truer appreciation of the physical earth. Chapter 10 thus presents an overview of analyses drawn from modern natural science on this topic. Suffice it to say, a broad consensus of the professional scientific community considers that much of the evidence referenced by flat earth proponents rather more authentically depicts the earth as a globe. It is useful to be aware of their reasoning, whatever one's position may be.

Flat earth perspectives have become sufficiently widespread within contemporary culture to attract the attention of modern cosmologists such as Neil deGrasse Tyson and Stephen Hawking. Both have responded in public rebuttals.[30] Since modern flat earth arguments challenge authentic Purāṇic discourse as well, this presentation will argue that while Purāṇic cosmology no doubt describes numerous features that challenge standard empirical experience, nonetheless, Purāṇic commentators dating back to classical antiquity also acknowledged observable phenomena that not only recognized the spherical nature of the earth of our experience, but mathematically quantified it. Evidence suggests that traditional Vedic commentators would not have seriously considered arguments readily identified with contemporary flat earth discourse.

For those who identify with this tradition, a central question remains as to how a discourse drawn from Purāṇic accounts containing seemingly fantastic cosmological phenomena in apparent conflict with observable astronomy acknowledged in traditional *jyotiṣa* accounts, can appear relevant to a contemporary audience. According to Ravi M. Gupta, the Charles Redd Chair of Religious Studies and Department Head of History at Utah State University:

> Discourses about [the nature of] reality have been present since the beginning of human conversation and certainly in the 16th century [at the time of Caitanya Mahāprabhu]. For example, with Jīva Gosvāmī, he was very well aware of the different philosophies, ideas, and scientific progress in his time. He had studied them well. He quotes from them, he engages with them; and engages in a way that he makes his tradition relevant to the discourse of his time ...

> We see the same thing is true later in the tradition with persons like Baladeva Vidyābhūṣaṇa in the 18th century. We see the same is true recently, in the modern period with persons like Bhaktivinoda Ṭhākura, Bhaktisiddhānta Sarasvatī Ṭhākura, and A. C. Bhaktivedanta Swami Prabhupāda. Each one was committed to engaging with the dominant discourse of their time.
>
> I think this is the mark of a healthy religious tradition. It is the mark of a vibrant theology, one that is not only preserving the past, but it is preserving the past by engaging in it in the present.[31]

An investigation of conflicts and correspondences between esoteric descriptions in Purāṇic accounts and the exoteric Sanskrit mathematics of observable *jyotiṣa* astronomy highlight the vibrant intellectual tradition intrinsic to the Vedic cultural legacy. In contrast to certain schools of nondualist thought, the Gaudiya Vaishnava school acknowledges the tangible, yet temporary, quality of the material cosmic manifestation with full complexity. It also recognizes the many opportunities this discourse represents, by design, for practical intellectual engagement and genuine devotional contemplation.

While this work, *Vedic Cosmography in a Modern Context:* Virodha-parihāra *Revisited*, does not intend to resolve all concerns involving every aspect of the Purāṇic model, it will highlight some of the complexity that has engaged these issues in historical exchanges that continue into the present. There is more than sufficient evidence drawn from both classical and contemporary expressions of this tradition to justify that the spherical nature of the earth planet had been extensively acknowledged within a Vedic cultural context that also acknowledged Purāṇic accounts. While considering the earth as a central example, light can be shed on how the *ācāryas*, *sādhus*, and *paṇḍitas* of the past (and present), may have also handled the balance of Purāṇic discourse. That, arguably, is the main thrust of this project as an examination of how learned practitioners have maturely communicated content identified with the tradition.

SECTION I

Guru, Sādhu, and Śāstra

CHAPTER 1

Guru – Śrīla Prabhupāda on Fifth Canto Purāṇic Cosmography

a. "*Bhū-gola,* the earth is round ... *Bhū* means the earth; *gola* means round."

Śrīla Prabhupāda expressed unalloyed faith in the *Śrīmad-Bhāgavatam* while referring to the earth as a globe floating in an "ocean of air." This section examines multiple excerpts from Prabhupāda's morning walks, room conversations, and lectures that describe the globe-shaped nature of the earth and other objects located within a Purāṇic concept of space.

Prabhupāda frequently described Purāṇic features as round and/or egg-shaped, rather than flat. For example, in a 1975 *Bhāgavatam* lecture given in Nairobi, Africa, Prabhupāda clearly stated this preference:

Figure 3: Śrīla Prabhupāda sitting in his quarters in Geneva, Switzerland beside a globe of the earth.

> *Aṇḍa,* it is round. *Aṇḍa* means egg shape. We don't say it is flat ... *Śāstra* never says flat. Then you correct, "No, no, it is not flat. It is round." No, this is *śāstra.* Long, long years, in the beginning, it was called round, *aṇḍa.* Goloka, *go,* Goloka, Bhūrloka. *Goloka* means round. Goloka Vṛndāvana. Everything is round.[1]

Two years earlier, during a London *Bhagavad-gītā* class, Prabhupāda specifically mentioned that while some in the past may have considered a flat-earth model, Vedic *śāstra* has been describing spherical features for quite some time prior:

> **Prabhupāda:** *Bhū-gola. Gola* means round. I have heard that before the science, the people were under the impression that this world is square. Is it not?
>
> **Devotees:** Flat.
>
> **Prabhupāda:** Flat. Flat, yes. But in the Vedic *śāstra,* millions of years ago it is mentioned: *bhū-gola. Gola* means round. Just see.[2]

There were a number of times Prabhupāda directly described the earth planet as a "ball," as in the following example from one of his earliest *Bhagavad-gītā* lectures given in New York City: "Just like here, in this planet, when you go up, you see it is ball."[3]

Then again, in a discussion with his disciple Shyamasundar during the early 1970s involving a traditional example from the *Purāṇas*:

> And there is Varāha incarnation, He picked up the whole earth on the tusk. How much big the Varāha animal was to show that it can pick up the whole earth, earthly planet just like a ball.[4]

In the following series of statements, Prabhupāda specifically considered the motion of planets routinely considered spherical, along with other celestial objects:

> And actually, scientifically, it is true that due to ... the sun planet all other planets are rotating; otherwise they'll fall down. But they're floating in the air, in the sky.[5]

> Just like the earth has its orbit – it is rotating – similarly, every planet is rotating.[6]

> Everything is moving. All these planets are moving. Your earthly planet is also moving – within twenty-four hours covering 25,000 miles, day and night ... Similarly, every planet is rotating.[7]

> [A]ll these planets, they are rotating with high speed. Just like this planet. It is rotating 25,000 miles in twelve hours. Is it not? The circumference of this earth is 25,000 miles and ... day and night, [it moves] twenty-four hours. Almost one thousand miles per hour.[8]

> [T]he earth is running at one thousand miles per hour, but we cannot understand. The arrangement is so nice. The perfection that it is ... we cannot understand it. It is practical. We see the morning, day, coming. That means earth is moving.[9]

Following this last example, Prabhupāda then offered as his conclusion:

> When the airplane moves also, there are so many jerking [sounds] ... But here you see that such perfect arrangement – it is moving one thousand miles per hour, and there is no jerking. There is nothing of the sort. We are thinking, "We are sitting in the same place." And there is no brain [behind this creation]?[10]

In another anecdote that Prabhupāda would share while referencing his global travels during a temple *Śrīmad-Bhāgavatam* class, he states:

> And again, if you start your plane in sunlight, in the morning, in daylight ... you will find ... always light. Sometimes

> we have got experience, we start from a place, say, at ten o'clock, and going western side. Then, after few hours, we see it is still ten o'clock, and the light is there. This is our practical experience.[11]

At the very least, Śrīla Prabhupāda did not appear to express reservations regarding the earth as a globe in motion while discussing his outreach work. In fact, he appeared very comfortable doing so. Arguably, all forms of śāstric analysis can be considered an exercise in outreach work, both for internal and external audiences.

In a 1973 morning walk conversation with Bhaktisvarūpa Dāmodara Mahārāja, a leading founder of the Bhaktivedanta Institute, Śrīla Prabhupāda offered the following explanation of the Purāṇic concept of the earth:

> Yes, world is round. That is fact. Goloka.
>
> In Vedic literature it is *bhū-gola*, *jagad-aṇḍa*. These words are there. We can see also it is round, *jagad-aṇḍa*. The universe is round. And Goloka, or *Bhū-gola*. *Bhū-gola*, the earth is round ... *Bhū* means the earth; *gola* means round. It is already there. And the geography's called, according to Sanskrit, it is called *Bhū-gola*. Long, long ago, before Galileo.[12]

Figure 4: Śrīla Prabhupāda during a recorded morning walk conversation held along Venice Beach, CA with his disciples in 1973. Svarūpa Dāmodara (Bhaktisvarūpa Dāmodara Svāmī) is on Prabhupāda's left holding a microphone. Brahmānanda Mahārāja is to his right with a *daṇḍa*.

Śrīla Prabhupāda not only discussed the earth with his disciples utilizing language commonly identified with a globe, but he also used similar language when writing to his Godbrothers. For example, a few years after establishing ISKCON in 1966, Prabhupāda sent a letter to his God-brother, Dr. Nagendra Bapu, commending him for encouraging his daughter and son-in-law to visit the fledgling ISKCON temple in New York City. Prabhupāda also personally encouraged Dr. Bapu to likewise assist the youthful Gaudiya Vaishnava mission taking root in Western lands. He wrote: "There is immense potency of preaching the philosophy of Krishna Consciousness under the guidance of Lord Chaitanya." Prabhupāda then added:

> Of course, as far as possible I am trying to spread this movement all over the world, and at the present moment my activities are prominent from Hamburg to Tokyo, a distance of 14,000 miles. I think the circumference of the whole earth is 25,000 miles. So this should be covered by some of our Godbrothers so that Lord Chaitanya's message may be properly executed.[13]

This letter not only helps illustrate the aspirations handed down by Bhaktisiddhānta Sarasvatī Gosvāmī to his disciples such as Prabhupāda and Dr. Nagendra Bapu for introducing Krishna Consciousness around the globe, but also suggests that considering the earth in terms of a globe was an example of the lexicon he bequeathed to his disciples for appreciating his mission.[14] Of note, Śrīla Bhaktisiddhānta Sarasvatī was also considered a well-established scholar of Sanskrit mathematics and *jyotiṣa* astronomy, ancient Vedic traditions seemed comfortable considering the earth as a sphere in relative motion with other celestial objects as part of an extensive and ancient Vedic tradition that included Purāṇic literature. At the same time, Śrīla Bhaktisiddhānta Sarasvatī Gosvāmī's published commentaries on the *Śrīmad-Bhāgavatam*'s Fifth Canto account of Bhū-maṇḍala proved foundational for Śrīla Prabhupāda's English translations on Purāṇic cosmology.[15]

Figure 5: Śrīla Bhaktisiddhānta Sarasvatī Gosvāmī (c. 1930)

The next section will describe Bhaktisiddhānta Sarasvatī Ṭhākura's experience as a *brāhmaṇa jyotiṣa paṇḍita*. This aspect of the life of a great Vaishnava *ācārya* is especially relevant, as Bhaktisiddhānta Sarasvatī's translations and commentaries are frequently referenced in this work as well, most notably in Chapters 3 and 6.

b. Bhaktisiddhānta Sarasvatī Ṭhākura

Reinvigorating traditional Sanskrit astronomy, while establishing contemporary Purāṇic commentary

Śrīla Prabhupāda's spiritual master, Śrīla Bhaktisiddhānta Sarasvatī Gosvāmī, was born in Puri in 1874 with the given name Bimala Prasāda. His father, Bhaktivinoda Ṭhākura, helped established the modern foundations for a worldwide *saṅga* of Gaudiya Vaishnavism. As a youth, Bhakisiddhānta attended meetings of Bhaktivinoda's early institutional efforts such as the Viśva Vaiṣṇava Rāja Sabhā (the Royal World Vaishnava Association) and the Vaishnava Depository research center. The research center included a press, which regularly produced the journal *Sajjana-toṣaṇī*. Bhaktisiddhānta would help his father proofread the magazine, among other activities. In turn, Bhaktisiddhānta Sarasvatī Ṭhākura later in life established the Gaudiya Math, a pan-India Vaishnava institution that made inroads pioneering an international footprint within the Western world.

As a young teenager, Bhaktisiddhānta attended the Calcutta Metropolitan Institution, where along with "the obligatory curriculum," Bimala Prasāda studied Sanskrit grammar and classical Hindu astronomy (*jyotiṣa*) with the local pundits. In his leisure time, he met with fellow students, with whom he discussed issues of culture, religion and society."[16] He was said to possess

> a precocious interest in mathematics and abstract thinking. He studied astronomy with pandits Mahesh Chandra Cudamani and Sundara Lala, who in 1889 awarded him the title "Siddhānta Sarasvatī" in honor of his proficiency in this subject. Soon thereafter he began to use the title on the front page of his own articles and books, something that added authority and legitimacy to his writing.[17]

Bhaktisiddhānta next entered Calcutta's prestigious Sanskrit College, where "apart from Sanskrit," he studied "mathematics, Indian philosophy, and ancient history." But in 1895, he entered a dispute with the headmaster who held the reputation as "one of the finest Sanskrit scholars in India," about the "precise location of the equinox, which was traditionally calculated from a fixed point of the zodiac." During the ensuing debate that turned into "a college-wide conflict," Bhaktisiddhānta concluded that "rather than compromise what he believed to be the correct interpretation of the data, it would be better to simply depart the college."[18]

At this point, in his early twenties, Bhaktisiddhānta established his own school, the Sarasvata Chatuspathi, in the home of his father. It provided astronomical tutorials preparing students for the entrance exams at Sanskrit College. In the meantime, Bhaktisiddhānta Sarasvatī "undertook the task of translating two classical astronomical texts from Sanskrit to Bengali, the *Sūrya-siddhānta* (1896) and the *Siddhānta-śiromaṇi* (1897), which he published with his own notes." He also produced "two astronomical periodicals, the *Bṛhaspati or the Scientific Indian* and the *Jyotirvida*."[19] He would use these journals to publish a number of Bengali translations of Sanskrit works that included, "*Bhauma-Siddhānta* according to western calculation," and the entire "*Ārya-Siddhānta* by Āryabhaṭa."[20] The Bhaktivedanta Research Center in Kolkata holds copies of many of these texts.

A few excerpts from Bhaktisiddhānta Sarasvatī's foreword to his translation of *Sūrya-siddhānta*, helps to reveal his mood:

> Western scholars ... marvel at the unlimited research found in the mathematically-based *jyotiṣa-śāstra* of the ancients of India. Among the eighteen most respected of such astronomical manuscripts ... *Śrī Sūrya-siddhānta* is unanimously accepted as the oldest. Negligence has pushed this book to the verge of oblivion. Indians have always been lackadaisical about protecting the unbroken glory of their ancestors, and I do not expect they will give that up upon the words of someone like myself.[21]

Bhaktisiddhānta added that, even though "at present the majority of persons reputed as learned are averse to *śāstrīya* understanding," he had presented his translation of the *Sūrya-siddhānta* "with the hope of attracting the attention of really intelligent thinkers toward the study of astronomy."[22] He further noted that while some may "follow only the chapter of Bhāskarācārya's *Siddhānta-śiromaṇi* that discusses mathematical calculations," he proposed that if by

> reading this original version they become submissive to their own gurus, to the predecessor gurus of Bhāskar and others, and to him who gave intelligence to them all, the *ṛṣi* who is the author of *Śrī Sūrya-siddhānta*, ... their internal burning pride may be extinguished.[23]

To be sure, as a young scholar, Bhaktisiddhānta Sarasvatī hardly appeared disinclined to pull punches while upholding the authenticity of the tradition, which in his view included more than just the math.

Ferdinando Sardella, an Associate Professor of the History of Religions at Stockholm University and a recent biographer of Bhaktisiddhānta Sarasvatī Ṭhākura, notes that even though Bhaktisiddhānta "received a modern Western education from public schools, of greater significance is the fact that he was also educated in traditional Indic knowledge." Sardella felt this "set him apart from his contemporaries. Few among his peers were eager to cultivate traditional knowledge, since it offered no tangible advantage in terms of securing a professional career."[24] Nonetheless, there are also accounts of this period describing Bhaktisiddhānta having been "offered a chair in the Calcutta University by Sir Asutosh Mukherjee, which he refused."[25]

Around the turn of the century, Bhaktisiddhānta Sarasvatī continued to feel "a strong calling for religious life" in spite of his "intellectual and academic interests, and the respectable level of success he had achieved in the areas of scholarship and occupation." He would meet Gaura Kiśora dāsa Bābājī, a "highly regarded Vaishnava ascetic and mystic," during a program at his father's home.[26] He subsequently became his formal disciple in 1901.[27] Sardella writes that "Bhaktisiddhanta did not have a specific conversion experience," as "throughout his life, he remained a Caitanya Vaishnava." Rather, "it was his social role and identity that underwent major transformations."[28] Though some would recollect that "even in childhood Bimala Prasāda was referred to by his spiritual preceptors as Bhaktisiddhānta Sarasvatī,"[29] Sardella reports that it was around the time of initiation that Bhaktivinoda formally "placed '*bhakti*' before '*siddhānta*' in his son's title, an award for his proficiency in reading and explaining Vaishnava texts."[30]

Bhaktisiddhānta Sarasvatī would later reflect upon the significance of his meeting with his spiritual master:

> When out of false ego I was thinking "I am a great scholar of mathematics and philosophy—let any big *paṇḍita* come at any time, day or night, and I will cut his proposition to pieces," at that time I got the *darśana* of the lotus feet of Śrī Gurudeva. He ignored everything that was previously appreciated in me—my truthfulness, my moral and pious life, and my intellectualism—knowing them to be of little value. When I saw that he ignored whatever was good in me, I realized how good he himself must be, who could neglect so many "good qualities" in me. What inconceivable wealth he possessed![31]

During this period, Bhaktisiddhānta Sarasvatī also published the *Śrī Navadvīpa Pañjikā* (1901) with "the help of the renowned *jyotiṣī* Nandulala Vidyasagara."[32] Bhakti Vikāsa Swami, another prominent biographer of Bhaktisiddhānta, writes about this work and how it proved to be

> the fulfilment of a desire of Śrīla Jagannātha dāsa Bābājī Mahārāja, who had apprised Śrīla Bhaktivinoda Ṭhākura of the need for an almanac specifically meant for persons cultivating *śuddha-bhakti*, for at the time the only almanacs available catered to *smārta*-dominated pantheistic Hinduism ... [In this edition] Śrī Siddhānta Sarasvatī promoted *harināma* by employing different names of Viṣṇu for each month, fortnight, and day. Additionally, he gave dates of Vaiṣṇava festivals, proper and improper timeframes for performing *saṁskāras*, and auspicious and inauspicious times of undertaking various activities.[33]

Bhaktisiddhānta would also produce the almanac, *Bhakti-bhāvana Pañjikā*, as well as another title published by his "school of astronomy," the *Almanac for the year 1917*.[34] Though Bhaktisiddhānta Sarasvatī no longer undertook *jyotiṣa* projects once accepting the role of *sannyāsa* and establishing the Gaudiya Math, he continued to cite "the *Sūrya-siddhānta* and the *Siddhānta-śiromani* several times in his *Anubhāṣya* commentary on the *Caitanya-caritāmṛta*."[35] Śrīla Prabhupāda likewise referenced these works in his translations of the *Śrī Caitanya-caritāmṛta*. One of the most significant is found in Prabhupāda's purport to *Adi-lila*, Chapter 3, text 8:

> These calculations are given in the authentic astronomy book known as the *Sūrya-siddhānta*. This book was compiled by the great professor of astronomy and mathematics Bimal Prasād Datta, later known as Bhaktisiddhānta Sarasvatī Gosvāmī, who was our merciful spiritual master. He was honored with the title Siddhānta Sarasvatī for writing the *Sūrya-siddhānta*, and the title Gosvāmī Mahārāja was added when he accepted *sannyāsa*, the renounced order of life.

Perhaps it is not entirely coincidental that in his own commentary on Fifth Canto cosmology, Prabhupāda relied

> primarily on Bhaktisiddhānta's *Gauḍīya Bhāṣya*, a twelve-volume Bhāgavata edition in Bengali script that includes Sanskrit commentaries of Viśvanātha Cakravartī and Madhva, along with Bhaktisiddhānta's two Bengali commentaries. Prabhupāda cites the same texts and commentaries referenced by his guru and, starting from the fifth canto, includes translations of his guru's chapter summaries.[36]

Bhakti Vikāsa Swami describes the nature of this pivotal work at the heart of Prabhupāda's mission:

> In 1923, Śrīla Sarasvatī Ṭhākura established the Gaudiya Printing Works in Calcutta and began fascicular publishing of *Śrīmad-Bhāgavatam* with Bengali translation and commentary, a project that took twelve years to complete. His commentary was in two parts: *Ananta-gopāla-tathya*, selected relevant references from *Śrīmad-Bhāgavatam* itself

> and other scriptures, and from the glosses of previous *ācāryas*; and *Sindhu-vaibhava-vivṛti*, his own original contribution, which was restricted to the portions of *Bhāgavatam* dealing with classic philosophical expositions.[37]

It can be confidently said that Śrīla Prabhupāda's spiritual master, Śrīla Bhaktisiddhānta Sarasvatī Gosvāmī, not only reinvigorated the integrity of India's long Vedic tradition of Sanskrit astronomy and mathematics, but that he also literally provided the foundation for ISKCON's perspectives on Fifth Canto cosmology presented in the *Śrīmad-Bhāgavatam*, which features lengthy descriptions of the Bhū-maṇḍala plane. As a highly accomplished *jyotiṣa paṇḍita*, Bhaktisiddhanta profoundly appreciated the mathematical aspects of Sanskrit astronomy that considered the earth as a globe in relative motion with other celestial spheres. And it was his *Gauḍīya Bhāṣya Bhāgavatam* commentaries that provided the basis for Prabhupāda's purports on the subject of Purāṇic cosmology. Apparently, for Śrīla Bhaktisiddhānta Sarasvatī Gosvāmī, both perspectives were part of the same tradition.

c. Understanding the Purāṇic cosmological texts

Śrīla Prabhupāda engages the Bhaktivedanta Institute to make a model the Fifth Canto

Since numerous features of Purāṇic cosmography described in the Fifth Canto do not appear to correspond with how human beings normally experience the world, Śrīla Prabhupāda requested his "PhDs" to "collaborate and study" the Purāṇic texts to "make a working model of the universe":

> Now our PhDs must collaborate and study the Fifth Canto to make a model for building the Vedic Planetarium ... So now all you PhDs must carefully study the details of the Fifth Canto and make a working model of the universe. If we can explain the passing seasons, eclipses, phases of the moon, passing of day and night, etc., then it will be very powerful propaganda.[38]

Bhaktivedanta Institute outreach thus proved a central concern for Prabhupāda. In *Perspectives on Bhaktivedanta Institute* (1979), Bhaktisvarūpa Dāmodara Mahārāja reminisced about a conversation he had with Śrīla Prabhupāda in May 1977:

> It was very clear that His Divine Grace Śrīla Prabhupāda was constantly thinking about the Bhaktivedanta Institute and how it would introduce Kṛṣṇa consciousness into the academic community.[39]

Another founding member of the Bhaktivedanta Institute, Sadāpūta dāsa (Richard L. Thompson), embraced the challenge of producing a model of the Purāṇic cosmos in pursuance of these instructions. An early example of his research can be traced to a 1978 letter exchange with a graduate student from Southern Illinois University who presented numerous questions involving Purāṇic cosmographical models. Sadāpūta responded by comparing descriptions offered by two different sources from within the tradition:

> [T]he *Sūrya-siddhānta* describes the Moon as being closer than the Sun, whereas the *Śrīmad-Bhāgavatam* states that the Moon is above the Sun. This apparent contradiction is quickly resolved when one realizes that "above" is not being used in the *Bhāgavatam* in the sense of line of sight distance. In fact, the *Bhāgavatam* is evidently describing a vision of the universe which is radically different in many of its basic conceptions from the picture that is familiar to us. To us this vision tends to appear very strange in comparison to the system given in *Sūrya-siddhānta*, which more or less corresponds with what we would expect on the basis of current scientific theories. Yet both the *Bhāgavatam* and the *Sūrya-siddhānta* have been well known to past generations of acharyas.[40]

Sadāpūta's lengthy correspondence with those similarly interested in this topic included thirty-three letters dated between 1977 and 1989, at which time the Bhaktivedanta Book Trust published his pioneering work, *Vedic Cosmography and Astronomy*. Alongside his many Bhaktivedanta

Institute research interests, Sadāpūta presented and published scholarly analyses on Purāṇic cosmography throughout his life. Numerous scholarly-minded devotees within ISKCON have subsequently likewise taken up this challenge.

CHAPTER 2

Sādhu - Insights from Previous *Ācāryas*

a. Statements from Fifth Canto commentaries

Several quotes from the previous ācāryas in our sampradāya

Śrīla Prabhupāda would draw upon sets of commentaries that contained works of numerous recognized *sādhus* and *ācāryas* as a reference for his own *Srimad-Bhāgavatam* commentaries. The following list offers a sample of statements taken from these works that suggest a comfort level amongst these commentators, using language suggesting a spherical nature for the earth while discussing Fifth Canto cosmology. The translations in this section are drawn from *Commentaries on 5th Canto Bhāgavatam Cosmology* by Vaishnava *ācāryas*, edited by Danavir Goswami.

Śrīdhara Svāmī noted that while "narrating the story of King Priyavrata, mention was made of the divisions of the globe into islands" (SB 5.16.1). In his commentary for verse 5.25.12, Śrīdhara Svāmī also made mention of the "earthen sphere along with mountains." Meanwhile, Viśvanātha Cakravartī stated, "The earth rests like an atom on only one head [of Śeṣa]." Yādavārya would comment, "He [the Lord] bears the earth on His head like an atom."[1]

Regarding this topic, Prabhupāda would offer alternate *Bhāgavatam* translations within his purport to *Caitanya-caritāmṛta* (CC), *Ādi-līlā* 5.119. In them, he compares the "global spheres" situated on the hoods of Lord Śeṣa Nāga, to mustard seeds. Mustard seeds, of course, are distinctly spherical. Prabhupāda also writes in the CC purport his alternate translation for SB 5.25.2 – "Lord Anantadeva has thousands of hoods. Each sustains a global sphere that appears like a grain of mustard." Prabhupāda similarly writes in this purport an alternate translation for verse 5.17.21 – "On Your thousands of hoods rest the innumerable global spheres, like grains of mustard."

Among contributors to the traditional commentaries utilized by Śrīla Prabhupāda, Śrī Vaṁśīdhara Śarman has been described by one scholar as the "loquacious commentator on the *Bhāgavata*, generally occupying the second place in the list of commentaries just after Śridhara Svāmin's *Bhāvârthadīpikā*."[2] Śrīla Prabhupāda also mentioned Śrī Vaṁśīdhara in a letter to Sumati Morarjee, his early patron, as second on a list of "important commentaries on the *Bhāgavatam*."[3] Richard Thompson took special note, dedicating Appendix 1 in *Vedic Astronomy and Cosmography* to an analysis of Vaṁśīdhara's commentary on verse 5.20.38. For English translations of the Sanskrit text, Thompson secured assistance from two well-established Sanskrit scholars affiliated with important academic institutions in India.[4] Please see Appendix 4 for a copy of the full translation.

The following excerpt taken from these translations of Vaṁśīdhara's commentary for *Śrīmad-Bhāgavatam* 5.20.38 discusses his proposal for considering two different perspectives on the earth. Suffice it to say, they help illustrate the diversity of analytical approaches engaged in by respected commentators that can challenge a simplistic appreciation:

> In the Fifth Canto [5.25.2], "This great universe [*kṣiti-maṇḍalam* or earth-*maṇḍala*], situated on one of Lord Anantadeva's thousands of hoods, appears just like a white mustard seed." Thus by the illustration of a mustard seed it is known to be spherical.
>
> It is also said in the *Kardama-vihāra* [3.23.43], "After showing his wife the globe [*golam*] of the universe [*bhuvaḥ*] and its different arrangements, full of many wonders, the great *yogi* Kardama Muni returned to his own hermitage."
>
> In the Tenth Canto [10.8.37] it is stated, "She [Yaśodā] saw within His mouth all moving and nonmoving entities, outer space, and all directions, along with mountains, islands, the

surface of the earth [*bhū-golam*], the blowing wind, fire, the moon and the stars." By such proofs one should accept that there are two earths.

Well, this earth is the smaller one, so where is the other, bigger one?

Here is the answer: The bigger one is indeed a form of reflection [*pratibimba-rūpa*] up above the orbit of the asterisms. Its measure, according to SB 5.21.19 is, "the 95,100,000 *yojana* circumference of the sun's orbit around *bhū*."

[Commentary by Śrī Vaṁśīdhara on SB 5.20.38]

b. Śrī Vaṁśīdhara – Big Earth versus Small Earth

These comments by Śrī Vaṁśīdhara deserve a closer look. These statements were originally part of his *Bhāgavatam* commentary titled, *Bhāvārtha-dīpikā-prakāśa*. While scholars generally date it from the 19th century, a few had initially suggested the 17th century, and possibly earlier.[5] In this work, Vaṁśīdhara boldly discusses an enigmatic relationship between descriptions of an earthly realm and a massive Bhū-maṇḍala plane described in the Fifth Canto as billions of miles wide, and the small earth globe that corresponds with ordinary experience as described within the *jyotiṣa* tradition of Sanskrit mathematical astronomy.

Meanwhile, Śrī Vaṁśīdhara's analysis of the *Śrīmad-Bhāgavatam* also clearly expresses the finer sentiments of a well-versed Vaishnava. For example, in his *maṅgalācāraṇa* introducing his work, Vaṁśīdhara opens with the declaration:

> ... Obeisances unto Shri Ganesha!
> To Shri Krishna, the lover of Radha,
> Who goes around in the arbors in the Vrinda Forest,
> Who is surrounded by groups of cowherd girls
> —I take refuge in Him, the cause of everything.[6]

The subsequent excerpts from Śrī Vaṁśīdhara commentary on 5.20.38, were published in Appendix 1, "Vaṁśīdhara on Bhū-maṇḍala and the Earth Globe," of Thompson's *Vedic Cosmography and Astronomy*. In his analysis, Thompson described how Vaṁśīdhara "explicitly discusses the relationship between Bhū-maṇḍala, as described in the Fifth Canto, and the small earth globe of our experience." Thompson pointed out that Vaṁśīdhara's work helps to illustrate how the "astronomical literature known as *jyotiṣa śāstra* was accepted as valid by Vaiṣṇavas," while also considering "the apparent contradiction that exists between the cosmology of the Fifth Canto and this system of astronomy."[7]

Thompson continued: "Vaṁśīdhara tries to resolve this contradiction, and we should state clearly here that we do not think that his analysis is entirely correct. However, our own understanding is certainly far from perfect." He then added, "Thus it is best for us to carefully consider the information that is available to us and see what insights may gradually emerge," a topic Thompson explored throughout the appendix.

It is worth considering Thompson's analysis of Śrī Vaṁśīdhara's approach in light of some of the central issues discussed in this book. Thompson described how

> Vaṁśīdhara confronts the apparent conflict between the size of the earth given in the *Bhāgavatam* and the size given in the *jyotiṣa śāstra*. From *jyotiṣa śāstra*, he cites a value of 4,967 *yojanas* for the circumference of the earth globe. In fact, this figure is given in verse 3.52 of the *Siddhānta-śiromaṇi* of Bhāskarācārya, along with the value of 1,581 1/24 *yojanas* for the earth's diameter ... This agrees closely with our present figure for the circumference of the earth, using 5 miles per *yojana*. Vaṁśīdhara does not indicate the he thinks this figure is wrong. Rather, he accepts it without question and suggests various ways of reconciling it with Puranic cosmology.[8]

For example, in one passage Vaṁśīdhara wrote:

> Well, then how can one explain the contradiction between the *Bhāgavata* and the *jyotiṣa śāstra*, or astronomical science? In the *Bhāgavata*, Jambūdvīpa is said to measure

> 100,000 *yojanas*, whereas the astronomical science states the entire earth to be measuring only 5,000 *yojanas*.[9]

Vaṁśīdhara then offered a solution:

> The earth has two forms. One is the particular [*viśeṣa*] form of big measure, and the other is the smaller, spherical form given in the *jyotiṣa śāstra* … In the *jyotiṣa śāstra* the word *bhūgola* refers to the earth as an egg of clay, and the word *viśeṣa*, or "particular," refers to the round golden egg described in the *Purāṇas* … In accordance with the *Purāṇas*, Śukadeva Gosvāmī has spoken of the big measure of the earth, and only suggested the small measure.[10]

Thompson described in his Appendix that Vaṁśīdhara's commentary illustrated the "apparent contradiction between the Purāṇic cosmology represented by the Fifth Canto and the *jyotiṣa śāstra*," as considered by Vaishnava commentators. Thompson noted how the "*jyotiṣa śāstra* was regarded as the 'eye of the *Vedas*,' and it seems to correspond to observable reality. Yet its description of the earth seems totally at variance with the 'big earth' of 500 million *yojanas* described in the Fifth Canto."[11] Comments like these suggest an issue well-acknowledged within the Vaishnava theological tradition.

Thompson offered as a conclusion:

> We would suggest that "the big earth" [proposed by Vaṁśīdhara] corresponds to the reality directly perceived by persons on the level of consciousness of Śukadeva Gosvāmī, while "the small earth" corresponds to the reality perceived at an ordinary level of human consciousness. The two earths are both aspects of one underlying reality, and the relation between them is higher-dimensional: it cannot be understood in terms of the bending of light in ordinary, three-dimensional space. According to this idea, both earths are reflections, in an abstract sense, of the underlying reality.[12]

CHAPTER 3

Śāstra – Digging Deeper

a. Is the *Śrīmad-Bhāgavatam* compatible with "Earth as a Globe"?

Although the *Bhāgavata Purāṇa* does not directly refer to the earth as a small globe floating in space, the Fifth Canto does acknowledge structural elements compatible with a more limited spherical experience of the earth. In fact, the *Bhāgavatam* supports this relativity of conditional perception in two critical, indirect ways.

First, there are verses that appear to imply a spherical earth by referring to our planet as *bhū-gola, bhū-golaka,* or *bhū-golaṁ.*[1] For example, Śrīla Prabhupāda frequently translated the term *bhū-gola* in English as "round earth." Further, the famed Monier-Williams Sanskrit dictionary presents the definition for *bhū-gola*: "'earth-ball', the terrestrial globe, earth;" and for the term *bhū-golaka*: "the terrestrial globe."[2] The Sanskrit and Indian Studies scholar, Dr. Howard Resnick (Hridayananda das Goswami), who worked on completing the translation and commentary for the Bhaktivedanta Book Trust editions of the *Bhāgavata Purāna*, notes that as a standard Sanskrit-English translation, "*Gola* means 'globe, (as the celestial globe or as the globe of the sun or of the earth).' Similarly, *golaka* means 'a ball or globe, specifically in the *Śrīmad-Bhāgavatam*.'"[3]

As noted, Prabhupāda at times described: "*Bhū-gola*, the earth is round." But in another example from the Tenth Canto when Kṛṣṇa's friends complain to mother Yaśodā that Kṛṣṇa had eaten dirt, Prabhupāda translated the Sanskrit term *bhū-golam* as the "surface of the earth."

> When Kṛṣṇa opened His mouth wide by the order of mother Yaśodā, she saw within His mouth all moving and non-moving entities, outer space, and all directions, along with mountains, islands, oceans, the surface of the earth, the blowing wind, fire, the moon and the stars.
>
> [SB 10.8.37–39]

Elsewhere, Prabhupāda would also translate the Sanskrit term *bhuvo golam* as "globe of the universe":

> After showing his wife the globe of the universe and its different arrangements, full of many wonders, the great *yogī* Kardama Muni returned to his own hermitage.[4]
>
> [SB 3.23.43]

Yet Prabhupāda would expand on this description in his purport:

> All the planets are here described as *gola,* round. Every planet is round, and each planet is a different shelter, just like islands in the great ocean.

Another indication of accordance in the *Bhāgavatam* with the earth as a globe is based on the value of the *yojana*, a standard unit of length used throughout the text. While researching historical references to traditional units of measure found in the *Purāṇas*, Thompson uncovered an integral connection between the *yojana* and degrees of latitude, a spherical unit for measuring distance. Drawing upon his experience as a professional mathematician, Thompson reported that the *Sūrya-siddhānta* account for the length of a *yojana* appears consistent with an approximate length of five miles. Meanwhile, traditional scholars such as Parameshvara Nambudiri (c. 1380–1460) note that in the *Purāṇas*, the length of the *yojana* is approximately eight miles, as similarly utilized by Śrīla Prabhupāda and Śrīla Bhaktisiddhānta Sarasvatī Ṭhākura.

Interestingly, Thompson calculated that both lengths utilized in these traditional Vedic accounts could be "precisely defined as a fraction of a degree of latitude at the equator. Both relate to the earth by multiples of 108 (namely 432 and 1728), and this relationship gives us a very accurate estimate of the polar flattening of the earth." Thompson then restated that if using the length of the larger *yojana*, then specific measurements described in the Fifth Canto compared well with calculations identified with modern astronomy.[5]

Mathematical analyses such as the one just discussed suggest a compatibility between a spherical earth model as found in the *Sūrya-siddhānta*, and descriptions found in the *Bhāgavata Purāṇa*. The fundamental unit for measuring distance in both of these traditional Vedic texts, the *yojana*, is based on a spherical system of measurement (degrees of latitude).

b. The *Sūrya-siddhānta*

The *Sūrya-siddhānta* is a centrally important *jyotiṣa* treatise held in high regard by Prabhupāda's spiritual master, Śrīla Bhaktisiddhānta Sarasvatī Ṭhākura, who published a translation during his early days as a Sanskrit *paṇḍita*. Bhaktisiddhānta considered this work the "oldest and [most] universally accepted amongst the 18 basic texts of astronomy (*siddhāntas*) present in ancient India," and a "prime jewel of mathematics."[6] It is also a text widely recognized within the Gaudiya Vaishnava tradition.

Though it considers celestial motion from a geocentric perspective, its calculations agree remarkably well with certain aspects contemporary astronomy. They provide accurate predictions for the complex motion of the sun, moon, and planets as well as correct times for eclipses, solstices, and equinoxes. In addition, the *Sūrya-siddhānta* frequently describes the earth as a sphere, and offers relatively accurate dimensions not only for the earth planet, but other planets as well.

In Chapter 12 of Bhaktisiddhānta Sarasvatī Ṭhākura's translation, the *Sūrya-siddhānta* presents an explanation for why the earth as a globe may appear flat to some, like a circular disk:

> As the shape of the earth is spherical, people in each place think themselves to be uppermost. But as the globe of earth is situated in ether, what then is its upper and lower side? Because of the smallness of their size, people see this earth as circular in all four directions.
>
> [*Sūrya-siddhānta* 12.53–54]

In a classic 1860 English edition of the same text, the translator, the American Reverend Ebenezer Burgess, offered a slightly different phrasing for the second verse:

> Owing to the littleness of their own bodies, men, looking in every direction from the position they occupy, behold this earth, although it is globular, as having the form of a wheel.

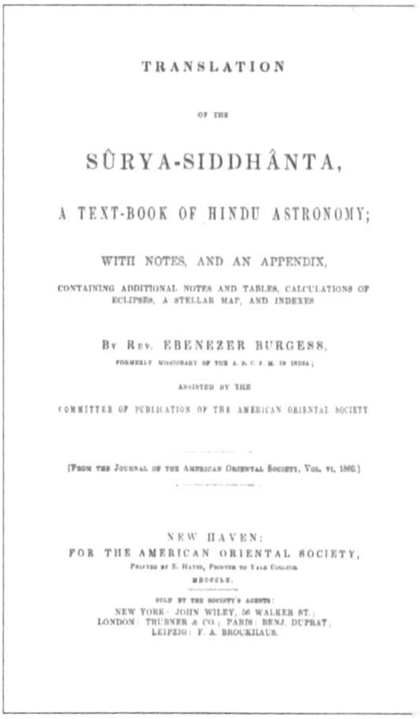
TRANSLATION

OF THE

SÛRYA-SIDDHÂNTA,

A TEXT-BOOK OF HINDU ASTRONOMY;

WITH NOTES, AND AN APPENDIX,

CONTAINING ADDITIONAL NOTES AND TABLES, CALCULATIONS OF ECLIPSES, A STELLAR MAP, AND INDEXES

BY REV. EBENEZER BURGESS,

FORMERLY MISSIONARY OF THE A. B. C. F. M. IN INDIA;

ASSISTED BY THE

COMMITTEE OF PUBLICATION OF THE AMERICAN ORIENTAL SOCIETY

[FROM THE JOURNAL OF THE AMERICAN ORIENTAL SOCIETY, VOL. VI, 1860.]

NEW HAVEN:

FOR THE AMERICAN ORIENTAL SOCIETY,

PRINTED BY E. HAYES, PRINTER TO YALE COLLEGE.

MDCCCLX.

SOLD BY THE SOCIETY'S AGENTS:

NEW YORK: JOHN WILEY, 56 WALKER ST.;

LONDON: TRÜBNER & CO.; PARIS: BENJ. DUPRAT;

LEIPZIG: F. A. BROCKHAUS.

Explanations such as these sound reminiscent of a comment written by Śrīla Prabhupāda in a letter to his disciple, Mādhava dāsa (Michael Marchetti), one of the founding members of the Bhaktivedanta Institute:

> Some portion of the earth is flat. When you stand in any place you see flat, so for us to some extent it appears flat, but it is round.[7]

Other verses from the *Sūrya-siddhānta* translated by Bhaktisiddhānta Sarasvatī also describe the earth as a globe:

> By dint of Brahmā's energy derived from (his) self-concentration, the earth globe (*Bhūgola*) is situated at the middle of the all spaces (ether—*Vyoma*) inside the egg (*Brahmāṇḍa*).
>
> [*Sūrya-siddhānta* 12.32]

> East of the fourth quadrant of the earth-globe is *bhadrāśva-varṣa*, and in the middle of that is *yamakoṭipurī*.
>
> Note: Burgess would phrase "earth-globe" as "earth's circumference" [*Sūrya-siddhānta* 12.38]

The famed *jyotiṣa* text also offers a number of comparatively accurate planetary measurements, according to a modern estimation. With specific reference to the verses just discussed, Antardwīp dāsa, a lead designer for the Temple of the Vedic Planetarium chandelier model, has written a paper offering a mathematical analysis of the spherical nature of the earth drawn from verses 12.38–41 in the *Sūrya-siddhānta*.[8] Meanwhile, in a 1997 paper titled "Planetary Diameters in the *Sūrya-siddhānta*," published in the *Journal of Scientific Exploration*, Richard L. Thompson presented a mathematical analysis of measurements found in the *Sūrya-siddhānta* for the earth's diameter. For example:

> The earth's diameter (*bhukarṇa*) is 1600 (2 × 800) *yojanas*. The square root of 10 times the square of earth's diameter is the earth's circumference.
>
> [*Sūrya-siddhānta* 1.59][9]

Thompson described in his essay: "The diameters are measured in a unit of distance called the *yojana*, which in the *Sūrya-siddhānta* is about five miles." Since verse 1.59 gives the figure of 1,600 *yojanas*, that would yield "8,000 miles as the approximate diameter of the earth." Modern calculations offer 7,926 miles (12,760 km). As for the circumference of the earth, the *Sūrya-siddhānta* describes the equivalent of 25,300 miles (~40,700 km), whereas the modern figure is 24,901 miles (40,075 km). Thompson also notes that the *Sūrya-siddhānta* "gives explicit values" for the diameter of the moon as 480 *yojanas*, in verse 4.1. This figure corresponds to 2400 miles (~3900 km), whereas the modern calculation is 2159 miles (~3475 km).[10]

Particularly when considering the earth's dimensions, various traditional accounts can appear surprisingly in tune with contemporary calculations. Perhaps this could be taken with all the more intrigue while considering how these calculations were propagated within a cultural context that simultaneously upheld a variety of starkly divergent Purāṇic accounts. Suffice it to say, these particular mathematical calculations, disseminated within a broad tradition that included literature such as the *Sūrya-siddhānta*, offer descriptions of the earthly realm perceived as a globe.

Although some may hold an impression of the Vedic worldview essentially in conflict with contemporary scientific data, that is not the case. Historical and śāstric evidence clearly illustrate a well-established ancient legacy identified with Vedic culture that would negotiate numerous apparent distinctions between Purāṇic and *jyotiṣa* accounts. Indeed, as previously noted, recognized Vaishnava commentators such as Śrī Vaṁśīdhara have directly addressed a variety of apparent dichotomies, such as descriptions of a big earth/small earth involving a mathematically quantifiable globe versus an expansive cosmological plane, in their Fifth Canto commentaries. Perhaps rather than being an anomalous activity meant to be sidelined, this sort of enterprise represents a dynamic legacy central to the Vedic cultural tradition. Śrīla Bhaktisiddhānta Sarasvatī Ṭhākura appeared to consider it thus, what with his extensive critical analysis of Sanskrit mathematical astronomy along with his substantive commentary on Purāṇic cosmography described in the Fifth Canto.

SECTION II

Purāṇic and *Jyotiṣa* Accounts

CHAPTER 4

The Legacy of *Jyotir Veda*: The Tradition Speaks for Itself

Bhānu Swami writes in the Foreword to *Commentaries on 5th Canto Bhāgavatam Cosmology*:

> The structure of the universe and calculations of planetary movements have been a matter of great interest to many civilizations. The Vedic scriptures give ample evidence of this interest. The ancient calculations give similar results to modern methods, whereas the Vedic description [offered in the *Purāṇas*] of the structure of the universe differs greatly.[1]

Correspondingly, the author of *Vedic Cosmography and Astronomy* (BBT 1989), Richard L. Thompson (Sadāpūta dāsa) reported in a seminar given on this topic in 1986:

> This apparent conflict in understanding the Fifth Canto [of the *Bhāgavata Purāṇa*] is not something new. And it is not a conflict between Vedic literature and modern scientific understanding. Actually, there is a whole body of Vedic literature which describes the earth as a small globe. And this literature is what Śrīla Prabhupāda in his purports refers to as *Jyotir Veda* or *Jyotiḥśāstra*.[2]

Thompson proposed that "to understand the Fifth Canto ... you also have to understand the *Jyotiḥśāstra*, because *Jyotiḥśāstra* and the Fifth Canto fit together as part of one unified tradition."[3]

This section will present an introductory exploration of the long conversation that took place within a well-established cultural tradition that identified with Vedic literature. It will specifically focus on discussions of the earth as a spherical object. Insights into the lives of some of the personalities involved, many well known in the world history of mathematics and astronomy, will be considered. In addition, it will present a glimpse of the professional scholarship that currently engages this topic. It is possible that many aspects of contemporary discourse involving dynamics that Prabhupāda has described as both "esoteric and exoteric,"[4] have proven essential to this ancient legacy.

a. The nature of a discourse spanning millennia

In his award winning work, *Hindu Theology and Biology: The Bhāgavat Purāṇa and Contemporary Theory*, Jonathan Edelmann describes "a large body of literature ranging from approximately AD 500 to 1900 [that] examined the contradictions between the cosmology of the *Purāṇas* and Indian astronomical sciences contained in the Siddhāntas, Jyotiḥśāstra, and Tantras."[5] He reports that the *paṇḍitas* within this tradition acknowledged in their descriptions of the cosmos that "the *Purāṇas* provide no mathematical, predictive, or computation argumentation," and thus had "not been seen ... as a source of information with regard to calendar making, astronomical predictions, [and] astrological evaluations." Rather, for "practical purposes," texts like the *jyotiṣa siddhāntas* were consulted for "calculating time for the proper execution of sacrifice, discovering precise directionality for sacrifice, [and for] knowing the correct placement of stars, planets, etc. for ritual purposes."[6]

While various historically prominent astronomers and mathematicians intimately identified with traditional Indian culture – such as Āryabhaṭa (476–550 CE), Brahmagupta (598–c. 668 CE), and Bhāskara (1114–1185) – may not have utilized Purāṇic accounts for "empirical and mathematical" content, neither did they "reject ... [the] astronomical contributions of the Purāṇas in total."[7] As a result, efforts were made to settle differences as perceived by practitioners within the broad tradition. The Sanskrit term for this activity, *Virodha-parihāra*, translates as "removal of contradiction."[8] In many respects *Virodha-parihāra* discourse continues within a contemporary cultural context.

Oxford Professor of Sanskrit, Christopher Minkowski, describes how the "removal of contradiction can conceivably be done in a variety of ways." He reports that among them:

> Some ... authors simply refute one or the other of the contradictory views as untruth or illogic ... Others accord different levels of truth to the two positions, saying that one describes practical, the other, ultimate reality ... Some adopt an exegetical strategy, according one of the two views a higher truth ... Others assume that both views have the same subject and purpose on the same plane, and ... adjust and revise doctrines so that they conform to each other. Some ... resort to combining several of these theoretical approaches.[9]

Minkowski notes that for centuries prior to European contact this took shape as a "'standard' accommodation" model that placed an emphasis on *jyotiṣa* insights.[10] Coincidently, during the life of Caitanya Mahāprabhu, this previously dominant approach began to experience a variety of challenges from *brāhmaṇa jyotiṣa paṇḍitas* who favored heightening the relevance of Purāṇic perspectives.[11]

The main point in this section thus will stress that celebrated practitioners from India's *jyotiṣa* and Purāṇic traditions engaged within a historically verifiable discourse for well over a millennium and that further, they were well aware of the many issues that still appear to confound contemporary practitioners. These *paṇḍitas* and *ācāryas*, deeply immersed within a culture identified with Vedic literature, employed several intriguing modes of intellectual inquiry, philosophical analysis, and seasoned debate.

As for devotional perspectives specifically drawn from the tradition of Gaudiya Vaishnavism, Śrīla Prabhupāda concluded his commentary on Śukadeva Gosvāmī's introductory verse on Purāṇic cosmography (SB 5.16.4, Purport), as follows:

> The limits of the expansions of Govinda, the Supreme Personality of Godhead, cannot be estimated by anyone, even a person as perfect as Brahmā, not to speak of tiny scientists ... We should therefore be satisfied with the information

> obtainable from Vedic sources as spoken by authorities like Śukadeva Gosvāmī.

Along similar lines, the celebrated Vaishnava, Vīrarāghavācārya, offered the following account for the same verse. In his *Bhāgavata-candrikā* commentary (14th century), Śukadeva Gosvāmī speaks to Mahārāja Parīkṣit:

> as the forms of the Lord are infinite, it is impossible to deal with them exhaustively by mind or words. Still, I promise to explain in more detail the earth globe ... I shall dwell on the prominent details of the earth-globe, its divisions, including the names, extents and nature, etc., to answer your inquiry.[12]

b. Accounts of the *jyotiṣas* from ancient India

In the *Jyotiṣa-vedāṅga*, an ancient text that contemporary scholars assign a historical date from as late as the last centuries BCE to as early as 1400 BCE, the author, Lagadha, proclaims:

> The Vedas have indeed been revealed for the sake of the performance of the sacrifices. But these sacrifices are dependent on the (various segments of) time.[13]
>
> Therefore, he who knows *jyotiṣa*, this science of time, knows all.[14]

Varāhamihira (b. 505 CE), the celebrated astronomer, philosopher and mathematician from Ujjain, India, likewise offered a description of the role of the *brāhmaṇa jyotiṣa paṇḍitas* in *Brihat Samhita*, Chapter II, verses 14–16:

> That Brahmin Jyoutishaka, who has mastered both the text and the purport of the entire science, deserves to be respected and fed first on occasions of *Sradha*. He purifies the party of diners.

> Even the Mlechas and the Yavanas (Greeks), who have well studied the sciences, are respected as Rishis. Such being the case, if the Jyoutishaka should happen to be a Brahmin, who will deny him respect?
>
> He who, not having studied the science, [and who] passes for a Jyoutishaka, is a sinner and a disgrace to society.[15]

By all appearances, the tradition holds in esteem the *jyotiṣa paṇḍitas* and their studies. Of special note is Varāhamihira's statement that anyone, from anywhere, if proven "well-studied" in these "sciences," is meant to be "respected" as a Rishi. Conversely, those who pose as pretenders, no matter the tradition they claim to identify with, are considered "a sinner and a disgrace to society." Without doubt, the astral sciences were not a subject taken lightly.

This essay will next examine the lives a few outstanding *jyotiṣa paṇḍitas*, many of whom are well known within the global history of astronomy and mathematics. In particular, their thoughts will be considered on the nature of a spherical earth.

CHAPTER 5

Historical Introduction to India's *Jyotiṣa* Tradition

As for leading *brāhmaṇa jyotiṣa paṇḍitas* of mathematics and astronomy profoundly identified with South Asian culture, the examples of Āryabhaṭa (b. 476), Brahmagupta (b. 598), and Bhāskarācārya (b. 1114), can be taken as illustrative of classical discourse. This section will also examine another noted mathematical astronomer, Jñānarāja, a contemporary of Caitanya Mahāprabhu. While all were elite *paṇḍitas*, they by no means represented all views. Having said that, they held a general consensus considering the earth a sphere while upholding an intimate allegiance profoundly respectful of the Vedic cultural tradition.

The intended point: classical discourse identified with the Indian subcontinent proved familiar with numerous constructs frequently recognized as "modern." Furthermore, this discourse took place for well over a millennium prior to the rise of the "new sciences" identified with a European origin. While this short essay is not intended to offer an exhaustive survey of all concerns, it will provide illustrative examples of *jyotiṣa* views of the earth as a perceivable globe.

a. Āryabhaṭa (476–550 CE)

As described by historian of mathematics, Radha Charan Gupta, Āryabhaṭa's legacy endures as

> a celebrated astronomer and mathematician of the classical period of the Gupta dynasty (AD 320 to ca. 600). This

> era is called the Golden Age in the history of India, during which ... Indian culture and civilization reached a unique stage of development which left its deep impression upon succeeding ages.[1]

The astronomer and physicist, Indian historian of science S. M. Razaulah Ansari, tells of an "ancient Sanskrit saying" that reflects upon Āryabhaṭa's reputation. He writes of a verse that proclaims:

> When the methods of the five Siddhāntas began to yield results conflicting with the observed phenomena such as the setting of the planets and the eclipses, etc., there appeared in the *Kāli* age at Kusumapura *Sūrya* himself in the guise of Āryabhaṭa, the *kulapa* well-versed in astronomy.[2]

Āryabhaṭa acknowledged the source of his learning, "By the grace of Brahmā."[3] He ascribed his work "as the "universally true science of astronomy, which formerly was revealed by Svayambhu, and is now described by me in this *Aryabhatiya*."[4]

Figure 6: Statue of Āryabhaṭa at the Inter-University Centre for Astronomy and Astrophysics in Pune, India.

Āryabhaṭa would portray the earth as a globe while drawing upon mathematical astronomy considered from a geocentric perspective. For example, in his *Āryabhaṭīya* (499 CE), verse 6 of Chapter 4 (titled "*Gola* or the Sphere"), Āryabhaṭa stated:

> The globe of the Earth stands ... in space at the centre of the circular frame of asterisms [constellations] ... surrounded by the orbits (of the planets); it is made up of water, earth, fire and air and is spherical (*lit.* circular on all sides).[5]

Āryabhaṭa would also describe the spherical earth rotating fully each day. He calculated the time involved for the earth's daily motion within microseconds of contemporary figures. Though most "astronomers of the *Siddhāntas* taught that the Earth is a sphere," a few would also subscribe to Āryabhaṭa's view of the daily rotation of the earth.[6]

b. Brahmagupta (598–c. 670 CE), and his commentator Pṛthūdaka Svāmīn (c. 830–c. 890 CE)

Brahmagupta earned a reputation for at times extending severe criticism toward his fellow astral *paṇḍitas* while upholding the integrity of the *jyotiṣa* tradition. In his own right, Brahmagupta became "one of the most celebrated mathematicians in the East," and someone "described by many modern historians as one of the most gifted mathematicians in the world."[7] For example, George Sarton, a leading founder of the modern discipline of the history of science, acknowledged Brahmagupta as "one of the greatest scientists of his race and the greatest of his time."[8] Among the mathematical achievements attributed to Brahmagupta is the "earliest treatment of zero in algebra," in his *Brāhma-sphuṭa-siddhānta* (628 CE).[9]

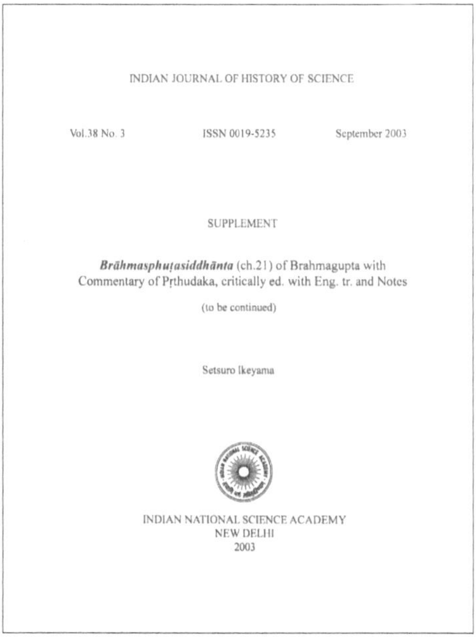
INDIAN JOURNAL OF HISTORY OF SCIENCE

Vol.38 No. 3 ISSN 0019-5235 September 2003

SUPPLEMENT

Brāhmasphuṭasiddhānta (ch.21) of Brahmagupta with Commentary of Pṛthudaka, critically ed. with Eng. tr. and Notes

(to be continued)

Setsuro Ikeyama

INDIAN NATIONAL SCIENCE ACADEMY
NEW DELHI
2003

In his introductory verses to this work, Brahmagupta offered acknowledgement to Brahmā and Śiva (Mahādeva), two predominating divine personalities recognized within a cultural context he intimately identified with:

> Mahādeva is victorious, whose feet are strewn with the splendor of the gems on the heads of bowing gods and asuras, and who causes the creation, endurance, and destruction of the world. (I, 1)

> The calculation of the planets mentioned by Brahmā, which became flaccid [weakened] after a long time, is explained truthfully by Brahmagupta the son of Jiṣṇu. (I, 2)[10]

While Brahmagupta's treatise examines numerous topics that continue to invite scholarly attention, for the purposes of this essay, discussion on perceptions of the earth as a globe is the focus. On this point, Brahmagupta is clear. For example, in his introduction to Chapter XXI, "Cosmological and Geographical Discussions":

> The rotation of the planets and nakśāstras is not the same everywhere for the residents of the earth. Because the knowledge about that is from the sphere, therefore I will explain the sphere. (XXI, 1)[11]

> The sphere of the earth which is surrounded by the spheres of the moon, Mercury, Venus, the sun, Mars, Jupiter, and Saturn, and which is inside the sphere of the stars, is shaped by the good and bad activities of creatures. (XXI, 2)[12]

In 864 CE, "the noted Sanskrit scholar" Caturveda Pṛthūdaka Svāmin, produced a commentary of Brahmagupta's work that "rendered the difficult slokas into simpler language."[13] Pṛthūdaka offered in his exegesis a description for XXI, 2, of the "true nature of the sphere such as the rotation of the planets or the shape of the earth":

> A sphere is a type of figure which is to be understood by all mathematicians since it is a figure. As grammarians obtain the correctness of a word by means of bases, affixes, (their)

> arguments, elision, sound changes, and (sound) augments, etc., ... in a like manner here calendar-makers obtain the true nature of the sphere such as the rotation of the planets or the shape of the earth by means of sines, arcs, versed sines, arms, uprights, hypotenuses, perpendiculars, rods, circles, etc., and by means of various types of true geometry.
>
> The roundness is determined by those who have destroyed contrary opinions just from the roundness of means of the characteristics which are to be found in a sphere; ...[14]

Pṛthūdaka Svāmīn included an assessment from one of Āryabhaṭa's first students in support of this argument.[15] He wrote: "Here the teacher Lāṭa (says) 'The sphere of the earth, symmetrically round, stands in heaven without support on all sides ...'"[16]

Pṛthūdaka then offered another verse, this time from the *Vasiṣṭha-siddhānta*. While traditional accounts attribute this work to the sage Vasiṣṭha eons prior to recorded history, contemporary scholars attest that it "appeared to have different versions and existed prior to the third century AD."[17] Caturveda Pṛthūdaka Svāmīn wrote: "So also in the *Vasiṣṭhasiddhānta*: The earth, consisting of the *mahābhūtas* (great elements) stands in the middle of the space of the cosmic egg for the existence of all creatures; it stands like a round sphere."[18] This verse, at the very least, locates discussions of the earth as a globe within a Vedic context, well into classical antiquity.

Brahmagupta's commentary toward the end of Chapter XXI in the *Brāhma-sphuṭa-siddhānta*, helps illustrate how the astral *paṇḍitas* negotiated exoteric mathematical data with the seemingly esoteric Purāṇic accounts within the context of a well-established cultural tradition. For example, Brahmagupta directly states while considering the shadow celestial entity, Rāhu, traditionally associated with eclipses: "therefore solar and lunar eclipses are not caused by Rāhu" (XXI, 38). But then a few stanzas later, in seeming contradiction Brahmagupta declares:

> Therefore the shadow of the earth does not cover the moon in an eclipse nor does the moon (cover) the sun; Rāhu which stands there and is equal (in size) to their diameters covers the sun and the moon. (XXI, 48)

How is it that both Brahmagupta and Pṛthūdaka Svāmin, his commentator, appear to endorse two apparently contradicting statements; one drawn from *jyotiṣa* mathematical astronomy, with the other identified with traditional Purāṇic accounts? Perhaps prior annotations will prove helpful. In them, Pṛthūdaka Svāmin offers contextual commentary:

> The statements of the itihāsas are intended to explain results of eclipses. Whoever causes the defectiveness of the sun and moon, that is rāhu for them. The statements in the saṃhitās of Garga and so on explain auspiciousness and inauspiciousness by means of signs such as the direction, color, and deflection (of eclipses). Whoever swallows (a luminary), that is also rāhu for them.
>
> However, while the true eclipser is not known, how can the time, magnitude, measure, duration, deflection etc. which agree with observation be known? Therefore whatever was mentioned before together with its explanation is correct.[19]

In these stanzas, Caturveda Pṛthūdaka Svāmin does not claim to have captured the ultimate cause when writing, "the true eclipser is not known." In this context, from his perspective, various accounts handle different aspects of a larger sense of cause and effect – some mathematically quantified, others perhaps less so. If considered this way, it would appear that the *jyotiṣa paṇḍitas* of the classical era could at times express an intriguing level of complexity coupled with intellectual dexterity, while grappling with seemingly disparate yet possibly parallel views of a grander cosmos held in the estimation of their audience.

CHAPTER 6

Bhāskarācārya (c. 1114–1185 CE), with Commentary from Bhaktisiddhānta Sarasvatī Ṭhākura (1874–1937)

In September 2014, an international conference of over 300 scholars gathered in Thane, India, to honor the 900th anniversary of Bhāskarācārya's birth. The proceedings, "appropriately named Bhāskara-prabhā (The Radiance of Bhāskara)," published a volume of papers in 2019. The editors, K. Ramasubramian, Takao Hayashi, and Clemency Montelle, describe in the work's introduction:

> Among the galaxy of mathematicians who flourished on the Indian subcontinent, Bhāskarācārya holds a very special place. Born in 1114 CE, Bhāskara wrote a collection of works on mathematics and astronomy which not only contained many technically brilliant results, but delighted readers with their poetic sophistication, imaginative appeal, and enduring charm. From the simple addition of two numbers, to elaborate algorithms accounting for the subtleties in planetary motion, Bhāskara enchanted his audience as much with his lucidity and eloquence, as his singular insight, intellectual ingenuity, and skills as a gifted teacher.[1]

In her book, *Mathematics in India* (2009), Kim Plofker similarly describes Bhāskarācārya's legacy:

> The closest thing to a universal canon of mathematical texts that emerged in Sanskrit was the output of the twelfth-century astronomer Bhāskara (II). His *Siddhānta-śiromaṇi*, *Līlāvatī*, and *Bīja-gaṇita* became by the middle of the second millennium the most widespread Indian expository works on mathematical astronomy, arithmetic, and algebra, respectively. Even today they are generally considered the "standard" structure for mathematical content.[2]

Figure 7: Statue of Bhāskarācārya in Patnadevi, India. Photo courtesy of Digviyay Sanjay Patil.

Perhaps without coincidence, Prabhupāda's spiritual master, Bhaktisiddhānta Sarasvatī Ṭhākura, keenly studied the works of Bhāskarācārya along with other notable *paṇḍitas* in the *jyotiṣa* tradition.[3] Bhaktisiddhānta Sarasvatī would also establish a school dedicated to Sanskrit mathematics and astronomy, prior to his taking *sannyāsa* and establishing the Gaudiya Math. During these years, Bhaktisiddhānta Sarasvatī published a translation with commentary on Bhāskarācārya's *Siddhānta-śiromaṇi*. He wrote in the introduction to the *Grahagaṇitādhyāya* section of *Siddhānta-śiromaṇi*:

> The special ability of *Śrī Bhāskarācārya* in mathematics, and also his ability to clearly and simply present even the most complex of theories in mathematics, makes him the undisputed leader amongst astronomers. The *Siddhānta-śiromaṇi*

of *Śrī Bhāskara* can be said to be an elaborate commentary on *Śrī Sūrya-siddhānta.*[4]

To secure a personal sense of philosophical outlook, examples from Bhāskarācārya's *maṅgalācāraṇa* invocation verses offer insight into his philosophical worldview. For instance, in the *Karaṇakutūhala* (1186 CE), a work whose title translates as the "Calculation of Astronomical Wonders," the author offers his respects:

> I, Bhāskara, having worshipped Lord Gaṇeśa, [Sarasvatī, the Goddess of] Speech, Brahmā, Viṣṇu, Maheśvara, and all the planets commencing from the Sun, venture to clearly set forth simple procedures for planetary computations which are compatible with the *Brahmasiddhānta.*[5]

Then in the *Golādhyāya* section of the *Siddhāntaśiromaṇi*, Bhāskara writes:

> Having paid obeisances to him (Ganeśa), by whose grace whatever is to be accomplished gets immediately accomplished, and likewise to her (Sarasvatī), by whose grace the speech of the writers dances on the stage of the tongue, filled with alluring words, various poetic embellishments, as well as playful elegance—like a well-adorned dancer, dancing with grace and elegance with beautiful feet-work—I state the spherics (*gola*) which is flawless and easily comprehensible to the uninitiated.[6]

Bhāskarācārya hailed from a brahmin family of accomplished *jyotiṣa* scholars who lived in a region ruled at the time by the Yādavas of Devagiri. During this era, the Yādava empire included much of present day Maharashtra along with portions of Madhya Pradesh and Karnataka. This royal lineage took its name from the famed Yādavas of the *Mahābhārata*, from whom they claimed descent.

While some scholars have considered the founder of the empire, Bhillama V, a patron of Bhāskara, historical records clearly chronicle Bhāskara's son, Lakṣmīdhara, as "the head of the learned men" at the court of the Yādava king Jaitrapāla. Meanwhile, Bhāskara's grandson,

Caṅgadeva, was "the chief astrologer" of the Yādava king Siṃghaṇa, while also serving as "the guru" of the local feudatory ruler "Soïdeva of the Nikumbha dynasty."[7]

Caṅgadeva established a *maṭha* described by scholars as a collegiate institution dedicated to the *jyotiṣa* sciences taught by Bhāskara. In the year 1207, Caṅgadeva installed a stone inscription describing the history and principles of the *maṭha*. The inscription clearly dedicated the *maṭha* "to spread the doctrines promulgated by the illustrious Bhāskarācārya."[8] The stone tablet additionally proclaims that the words of the great *jyotiṣa* "were rendered perspicuous by the enemy of Kaṁsa."[9] In other words, the *maṭha* attributed Bhāskarācārya's genius to Śri Kṛṣṇa, the famed "enemy of Kaṁsa."[10]

Figure 8: Caṅgadeva's inscription of 1207 at the Pāṭaṇādevī Temple in Patna, Bihar, India. Photo credit: K. Ramasubramanian, Devaraja Adiga, Aditya Kolachana, K. Mahesh. Image from: *Bhāskara-prabhā: Sources and Studies in the History of Mathematics and Physical Sciences* (Singapore: Springer, 2019), 6.

Specifically concerning Bhāskarācārya's greatest astronomical work, the *Siddhānta-śiromaṇi* (1150 CE), the theoretical physicist and historian of Indian mathematics, M. S. Sriram, describes it as

> one of the most comprehensive treatises on Indian astronomy. It has two parts, namely, *Grahagaṇitā* and *Golādhyāya*. *Grahagaṇitā* expounds on all the standard calculations and algorithms in astronomy of Bhāskara's times, whereas

> *Golādhyāya* had the definitions, more fundamental issues (like the nature of the earth, the placement of stars and planets around it and so on), and the principle and theoretical details of these calculations.[11]

While discussing texts such as Bhāskara's *Siddhānta-śiromaṇi*, Plofker explains that even though the "role of mathematics and mathematicians in premodern Indian society of any period is a potentially vast subject that is still poorly understood," historical evidence drawn from *jyotiṣa* manuscripts present numerous "claims" of the earth's "sphericity," suggesting that the concept was "universally accepted and stoutly defended in *siddhānta* texts."[12]

For example, in Bhaktisiddhānta Sarasvatī Ṭhākura's translation of *Siddhāntaśiromaṇi, Grahagaṇitā-dhyāya* verse 7.1, Bhāskara describes the spherical dimensions of the earth while considering how "the circumference of the earth is 4,967 *yojanas* and its diameter is 1,581 *yojanas*." These figures approximately correspond to 24,800 and 7,900 miles (~39,900 km and ~12,700 km), compared to modern figures averaging 24,900 and 7918 miles (~40,100 km and ~12,740 km).

Then Bhaktisiddhānta offers various alternate accounts taken from the tradition: "According to Āryabhaṭa, the circumference of the earth is 1050 yojanas ... According to the *Sri Sūrya-siddhānta* the circumference of the earth is 1,600." Bhaktisiddhānta Sarasvatī next adds another example identified with Western scholarship: "According to Prof. Beschel, the longest radius of the earth is 3962.802 miles and the shortest radius is 3949.555 miles."[13] Thus Bhaktisiddhānta not only acknowledges the spherical earth as discussed within a Vedic context, he felt comfortable utilizing Western resources.

Meanwhile, Bhāskara would bluntly state in verse III.2, that "the Globe of the Earth formed of ... Earth, Air, Water, the Ether, and Fire, is perfectly round."[14] In an effort to reconcile a spherical object appearing flat, Bhāskarācārya offered the argument:

> As the one-hundredth part of the circumference of a circle is (scarcely different from) a plane, and as the Earth is an excessively large body, and a man exceedingly small (in comparison), the whole visible portion of the Earth

> consequently appears to a man on its surface to be perfectly plane.[15]

Also of note, Bhaktisiddhānta Sarasvatī Ṭhākura's edition of the *Śrī Sūrya-siddhānta* states in verse 12.53: "the shape of the earth is spherical." Concerning the fine tuning of its spherical dimensions, verse 1.60 describes that the "circumference of the earth multiplied by the rSine of the co-latitude of a place and divided by the radius yields the corrected circumference of the earth."[16]

The point being, concepts of a perceivable and mathematically quantifiable spherical earth planet were well known amongst first-rate thinkers who identified with Vedic literature in the classical culture of the Indian subcontinent. Further, this discourse had been well established for centuries, if not for well over a millennium, prior to the development of the modern sciences identified with European origin.

CHAPTER 7

Jñānarāja (fl. 1503): Commentary at the Time of Caitanya Mahāprabhu

Jñānarāja authored the *Siddhāntasundara*, which translates as the "Treatise Beautiful." It was the first major *siddhānta* produced within the *jyotiṣa* tradition of Sanskrit mathematics and astronomy since Bhaksarācārya's *Siddhānta-śiromaṇi*, and it placed

> great emphasis on retaining and protecting Purāṇic authority ... with deep sympathies for and considerable knowledge of the astronomical sciences. [Jñānarāja] sought to show how the demonstrations, empirical data, and mathematics of the [Jyotiṣa] Siddhāntas could be reconciled with the Purāṇas, without the integrity or authority of either one being sacrificed."[1]

Coincidently, the year of its publication in 1503 also proved to be the same year Śrī Caitanya Mahāprabhu took initiation from Īśvara Puri, an event that subsequently launched Caitanya's pastimes propagating the *bhakti* tradition. Of note, Rūpa Gosvāmī, an intimate associate and leading proponent of the Caitanya *bhakti* movement known as a "renowned Gauḍīya Vaiṣṇava philosopher and scholar well-known to other *sampradāyas*," has also been considered "a master" of *jyotiṣa* "astrology and mathematics."[2] Caitanya Mahāprabhu's maternal grandfather, Nīlāmbara Cakravartī, was also respected as a highly accomplished *brāhmaṇa jyotiṣa paṇḍita*.[3]

As for Jñānarāja's family, his two sons, Cintāmaṇi and Sūryadāsa, were noted *jyotiṣa* scholars as well.[4] Their family traced its brahminical lineage back to the Vedic sage Bhāradvāja, the father of Droṇācārya, one of the central characters in the *Mahābhārata*.[5] Various manuscripts written by Jñānarāja's family suggest that they held pronounced Vaishnava sympathies. Of note, Jñānarāja's son, Sūryadāsa, is recognized for his commentaries on *śāstra* such as the *Bhagavad-gītā*, as well as for his Sanskrit poetry on similar topics. He is also credited for inventing a form Sanskrit bidirectional poetry (*vilomakāvya*), "in which a poem is composed in such a way that it can be read meaningfully from left to right, and from right to left, and in which two different narratives are sustained in the two directions." Sūryadāsa work, *Rāmakṛṣṇavilomakāvya* (c. 1580), describes the pastimes of Rāmacandra when read from one direction, and those of Kṛṣṇa when read from the other.[6]

Concerning Jñānarāja's *Siddhāntasundara*, it is noted for its "much stronger Purāṇicizing tendency, coupled with a greater concern with artistic expression."[7] Of special note is his chapter describing the seasons, a feature to which Bhāskarācārya also dedicated a chapter in the *Siddhānta-śiromaṇi*. But in Jñānarāja's version, twice as long as Bhāskara's, pastimes of Kṛṣṇa in Vṛndāvana forest frame the discussion of spring and summer, while descriptions of the autumn season features pastimes of Rāmacandra and Sītā.

Christopher Minkowski, a leading scholar on the tradition of accommodation and reconciliation between India's *jyotiṣa* and Purāṇic cosmological legacies, describes Jñānarāja's *Siddhāntasundara* as

> the first astronomical Siddhānta, or magisterial treatise, to appear in Sanskrit in 350 years ... To write another Siddhānta after a long gap is to recuperate the past, but at the same time to break from it, and the Siddhāntasundara pointedly defines itself by its departures from Bhāskara's work. Jñānarāja's work was circulated widely in north India, and defined the terms of a number of discussions that took place among the astronomers in Banaras [Varanasi] in the following century.[8]

Toke Knudsen, who translated Jñānarāja's *Siddhāntasunara* into English, writes of Jñānarāja's fresh approach:

> Traditionally, the approach of the Indian astronomers to the

> problem of the two cosmologies contradicting each other was to incorporate certain elements of the cosmology of the *purāṇas* into their model while rejecting others ... [In contrast,] Jñānarāja's approach was to seek a synthesis between the cosmology of the *purāṇas* and the cosmology of the Indian astronomical tradition which preserves the authority of the *purāṇas* (and therefore does not reject elements from them) while still maintaining the basic cosmological model in order to have a workable science.[9]

Minkowski reports that Jñānarāja's chapter on cosmology

> displays a much greater concern with textual authority, of the sruti, and, especially, the Purāṇas. He breaks with Bhāskara's understanding on several points ... When it comes to the planets and stars above, he resorts to what might have been an old argument, namely that the Purāṇas and Siddhāntas are both true, but describe the world in different creations (kalpabheda), the Jyotiḥśāstra being pertinent to our experience today.[10]

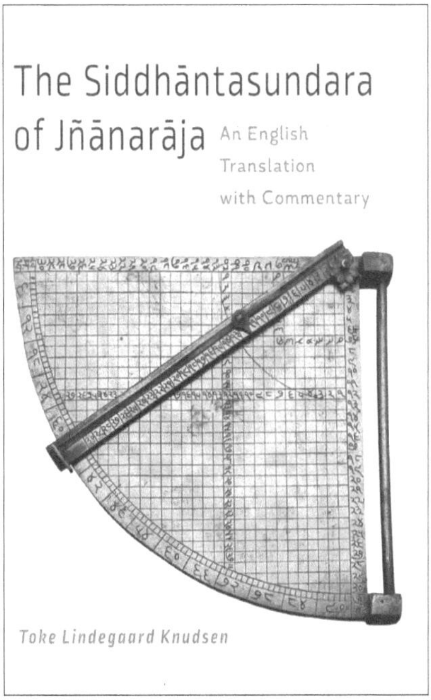

Along similar lines of reasoning that offered a "variant" on the "'different creations' argument," Jñānarāja's son, Sūryadāsa, utilized as "main evidence" in his *virodha-parihāra* chapter at the conclusion of *Siddhāntasaṃhitāsārasamuccaya*, the "Mucukunda story of the *Bhāgavata Purāṇa*, in which Mucukunda ... awakens from a long sleep" that stretched over countless years. Upon awakening, Mucukunda noticed that "everything and everyone has gotten smaller." While considering this account suggesting a massive change over deep time, Sūryadāsa argued that "the Purāṇas and Siddhāntas are both correct, but describe earlier and later stages in the history of creation."[11]

As for Jñānarāja, Minkowski describes that his chapter on cosmology concluded with "a statement of the purpose of cosmological description, which is to enable appreciation of the creation in all its multiplicity as nothing other than the Supreme Being." For example, Jñānarāja would explain:

> The one who knows the world as Vyāsa describes it achieves communion with the Deity. The purpose of cosmology—which is knowledge of the Creator's Creation—effectively eradicates the distinct purpose of Jyotis, since what is salvific is the *Bhāgavata's* account and not the Siddhānta's.[12]

While considering how during the years following Jñānarāja that various *jyotiṣa paṇḍitas* began to explore new initiatives of accommodation and reconciliation, Minkowski reports:

> All these works attempted in one way or another to demonstrate the non-contradiction between Purāṇic and Siddhantic models ... they [all] wrote primarily as defenders of Purāṇic authority and validity. The general approach of the argument is that faced with an apparent contradiction between Siddhānta and Purāṇa, the Purāṇa overrules the Siddhānta in its authority claim ...[13]

All the same, Jñānarāja directly states that the earth "is like a ball" in verse 1.24.[14] As previously noted, Jñānarāja bluntly explains in verse 1.27 that:

> [Even] after seeing the word *bhūgola* [that is, earth-sphere] being used in the *purāṇas* ... people who are disposed to obstinacy say that the entire earth is [flat] like the surface of a mirror. But none of them know the meaning of the *purāṇas*, nor that the spherical nature of the earth is well established by excellent demonstrations.[15]

Minkowski wonders why these efforts arose, which it could be noted, took place during a period shortly after Caitanya Mahāprabhu's birth. Minkowski describes that "so far no one has attempted an explanation of the appearance of the *avirodha* literature" that dealt with these issues. He does suggest that it "seems plausible to look to such historical factors as the rise of *Bhāgavata* worship and the towering importance of the Bhāgavata Purāṇa" that emerged during this period. He also mentions "the patronage by Rajput princes ... [and] of Braj / Mathura as a religious site," among the possible contributing factors inviting scholarly consideration.[16]

Jñānarāja and his sons helped establish a new tradition within *jyotiṣa* astronomy that both revisited and reinvigorated intellectual discourse on the relationship between Purāṇic cosmology and *jyotiṣa* astronomy. As previously noted, this essay does not attempt to offer a definitive overview of the breadth of concerns in play, what to speak of analyzing their merit. For this presentation, the emphasis rather touches upon a general awareness within an ancient scholarly tradition that considered their various apparent contradictions. Specifically, for this essay, attention revolves around concepts of a spherical earth well known well prior to the rise of European science.[17]

While upholding Purāṇic accounts in a manner not previously done within debates engaging the two traditions, Jñānarāja argued that perceptions of the earth as a sphere held true to Purāṇic understanding. Minkowski describes that according to Jñānarāja, the evidence "Pṛthūdaka gave for the sphericity of the earth was not accepted by those with base minds." As for those "who cite the authority of the Purāṇas to show that the earth is flat," Jñānarāja considered it, "Mere prating."[18]

As for the flavor of Jñānarāja's presentation, a few verses follow. Concerning to whom he attributed credit for his presentation, Jñānarāja writes in verse 1.3 of the *Siddhāntasundara*:

> The foremost knowledge concerning the nature of the motion of the planets and the stars that was related to Nārada by the four-faced [Brahmā] was composed in its entirety by the sage bearing the name Śākalya. In my own verses, I am presenting precisely that [knowledge], accompanied by demonstrations.[19]

And yet all the while, Jñānarāja maintained refined Vaishnava sentiments as expressed in his Chapter 6 describing the seasons. This section focused upon pastimes of Śri Kṛṣṇa and His Vṛndāvana associates during the spring and summer, followed by descriptions of Lord Rāmacandra during autumn. Included among the many examples of Jñānarāja describing Kṛṣṇa in Vṛndāvana:

When sweet fortunate Kṛṣṇa comes, His beloved exults in the forest dwelling. Kṛṣṇa delights and greets goodly people with the music born from the flute that He fills with His breath.[20] (6.2)

[Kṛṣṇa], who is the source of all brilliance, drowns out [with His splendor] the light of the sun, moon, and stars. Beautifully blowing His flute, resounding, the divine one rains down ambrosia here.[21] (6.16)

Interesting to note how in the latter verse, Jñānarāja wrote that the "brilliance" of Śri Kṛṣṇa "drowns out" concern for "the light of the sun, moon, and stars," suggesting a profound sense of philosophical priority.

CHAPTER 8

The *Jyotiṣa* Legacy: Concluding Thoughts

For thousands of years followers of the Vedic tradition have been keenly aware of the apparent dichotomy between the exact science of observable astronomy whose accurate charting proved essential for numerous religious activities, as well as the esoteric cosmological sections of *Śrīmad-Bhāgavatam*'s Fifth Canto similarly described in other Purāṇic literature. As mentioned in the introduction, within this tradition were highly accomplished *jyotiṣa paṇḍitas* such as Sūryadāsa in the 1500s and Kevalarāma during the early 1700s, who wrote treatises specifically dedicated to the issue of *virodha-parihāra* in an effort to reconcile apparent contradictions considered from standard human perception.[1] As for the nature of the earth upon which we live, proponents of the Vedic tradition readily acknowledged disparities between Purāṇic accounts of a massive Bhū-maṇḍala plane containing a region of human habitation identifiable with earthly experience, and *jyotiṣa* accounts of a mathematically quantifiable earth sphere more familiar to a modern audience. This work has examined these accounts as a heuristic device useful for reflecting upon broader issues considered within this particular tradition.

In his pioneering book, *Vedic Cosmography and Astronomy*, Richard Thompson commented upon the analyses attributed to the famed *brāhmaṇa jyotiṣa paṇḍitas*, many of whose "own religious and social tradition was based on the *Purāṇas*." For example, Thompson noted how Bhāskarācārya "could not reconcile the relatively small diameter of the earth, which he deduced from simple measurements, with the immense magnitude attributed to the earth by the Paurāṇikas."[2] Thompson suggested that while

> historical astronomers such as Bhāskarācārya were carrying on a genuine Vedic tradition of astronomy, their understanding of Vedic cosmology was nonetheless imperfect. They did not understand the higher-dimensional nature of structures such as Bhū-maṇḍala, and they consequently focused their attention on those features of Vedic astronomy that can be readily understood in three-dimensional terms.[3]

As per his own approach to these subjects, Thompson wrote that "the long-standing perplexities" that have historically "attended the subject of Vedic cosmology" are "not irreconcilable." He presented the argument:

> The apparent contradictions can be resolved by developing a proper understanding of the nature of space, time, and matter, as described in the *Śrīmad-Bhāgavatam*, and a corresponding understanding of the Vedic approach to describing and thinking about reality.[4]

No doubt, there is much to consider within this ancient cultural legacy well aware of both empirical observations identified with Sanskrit mathematical analyses, and the seemingly otherworldly Purāṇic accounts offering descriptions apparently beyond the bounds of standard human perception. Within this time-honored Vedic tradition, descriptions of the earth as a spherical object proved an assessment held amongst *paṇḍitas* who also honored the Purāṇic tradition. An investigation of the conflicts and correspondences between esoteric descriptions found in Purāṇic accounts, and the exoteric Sanskrit mathematics of observable *jyotiṣa* astronomy, help to highlight the vibrant intellectual tradition associated with the Vedic cultural legacy.

SECTION III

Contemporary Discussions: East and West

CHAPTER 9

Models of Bhū-maṇḍala Active Within ISKCON Discourse

In a traditional Vedic cultural context, *jyotiṣa brāhmaṇa paṇḍitas* recognized the earth as a spherical object in a manner consistent with observable astronomy, along with Purāṇic descriptions of a Bhū-maṇḍala plane seemingly wholly at odds with standard human perception. Within this *jyotiṣa* astronomical tradition of ancient India, many would attempt to reconcile these two contrasting perspectives in a manner that modern scholars have identified as a "'standard' accommodation" approach that "removes ... contradiction."[1] Such efforts were featured in the works of *jyotiṣa brāhmaṇas* such as Bhāskarācārya, whose reputation Bhaktisiddhānta Sarasvatī Ṭhākura held in high esteem. Similar discussions are also located in Chapter 12 of the *Sūrya-siddhānta* that also appear to describe a portion of the Bhū-maṇḍala plane projected onto the earth globe, with Jambūdvīpa mapped to the Northern Hemisphere, the rings of Sapta-dvīpa mapped to the Southern Hemisphere, and a representation of Mt. Meru featured prominently at the North Pole. Since the time of Caitanya Mahāprabhu, additional methods of accommodation have been considered as well.[2]

In other words, initiatives exploring contrast, correspondence, and potential synergy between observable data and Purāṇic accounts have proven to be an established part of the tradition. The following section offers a quick survey of contemporary models under consideration by ISKCON *paṇḍitas* who identify with this ancient legacy. In good part, these efforts are inspired by Prabhupāda's directive to design models of Fifth Canto cosmology that can be displayed in exhibits. While the descriptions in this section do not offer a comprehensive analysis, they will present a short overview that invites further exploration.

a. Richard L. Thompson on Purāṇic cosmology and mathematical astronomy

Richard Thompson, also known by his devotional name Sadāpūta dāsa, held a PhD in Mathematics from Cornell University. He was one of the founding members of the Bhaktivedanta Institute who, along with Bhaktisvarūpa Dāmodara Mahārāja and others, signed its incorporation papers in 1976. Thompson spent over 30 years researching Vedic perspectives on natural science. Two books in particular, along with several BTG essays, provide a focus on Purāṇic cosmology. The first book, *Vedic Cosmography and Astronomy*, published by the Bhaktivedanta Book Trust in 1989, offers a comprehensive introduction to the topic. The second, *Mysteries of the Sacred Universe* (2000), expanded upon the themes raised in *Vedic Cosmography*, while exploring mathematical analyses offering correspondence between various descriptions of the *dvīpas* of Bhū-maṇḍala and the planetary orbits of the observable solar system.[3] Thompson proposed in *Mysteries* that although Bhū-maṇḍala "may look like a naive flat-earth model, careful study shows that the *Bhāgavatam* uses the earth disk of Bhū-maṇḍala to represent at least four different things." Included among these models that offer correspondence are:

(1) A stereographic polar projection of the earth globe. Reducing a 3-dimensional sphere to create a flat 2-dimensional map has been a method utilized since antiquity for mapping both the earth and the night sky. A planisphere and an astrolabe offer examples of this technique utilized in Western culture.

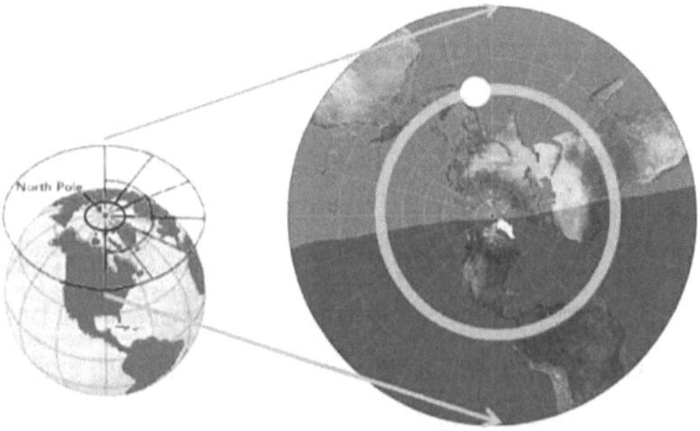

Figure 9: Planisphere as a stereographic polar projection of the earth globe. (Image used with permission from the Radha Mohan Das - Vedic Science and History Youtube Channel)

Figure 10: Mercator's astrolabe (1545), located in the Moravska Gallery, Brno, CZ.

In India, the famed globe of Mahārāja Sawai Jai Singh of Jaipur was an inverse projection of the plane of Bhū-maṇḍala onto a sphere produced during the early 1700s. Another interesting example of this kind of projection is the brass "Bhūgola or Earth-Ball, by Ksema Karna," dated 1571 and held in Oxford University's History of Science Museum. Several texts in the Fifth Canto describing Bhū-maṇḍala (such as 5.21.3, 5.21.8–9) are intriguingly suggestive of a planisphere model.

Figure 11: "Bhūgola or Earth-Ball by Ksema Karna, India, 1571" located in the Museum of the History of Science, University of Oxford.

(2) Bhū-maṇḍala as the ecliptic plane, which from a geocentric perspective corresponds to the plane of the sun's apparent orbit around the earth in three dimensions. Scholars such as Giorgio de Santillana and Hertha von Dechend have argued that the "flat earth" described by a variety of ancient cultures was also referring to the celestial plane of the ecliptic, with a depth of two dozen degrees latitude both north and south of this plane. Using this as a model, the planetary "heights" described in the Fifth Canto are shown to correlate with their orbital inclinations, and the mysterious "rings" of Bhū-maṇḍala mark the apogee and perigee of the geocentric orbits of the 5 planets, Mercury through Saturn.

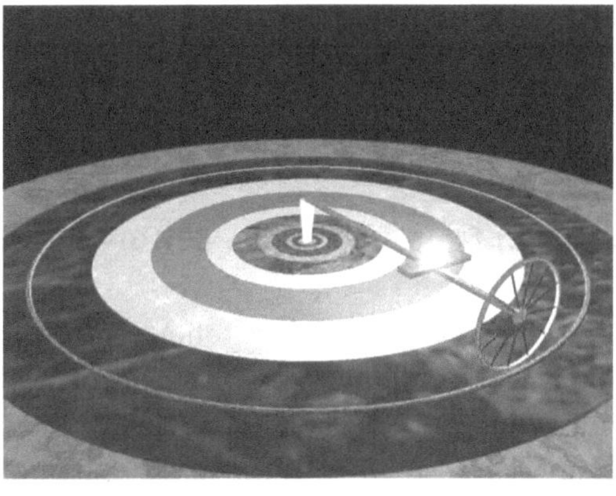

Figure 12: The rings of Bhū-maṇḍala as alternating islands and oceans, with a model of the sun's chariot (not to scale).

(3) Jambūdvīpa as a local map of greater India. Indologists have noted that many of the names for the geographical features in Jambudvīpa, the central island in Bhū-maṇḍala surrounding Mt. Meru, have counterparts in modern-day south-central Asia. At a reduced scale, various commonly named mountain ranges, rivers,

and countries consistently correspond with locations identified with Jambudvīpa.

(4) The celestial regions inhabited by demigods, or *bhaumāni svarga-padāni*. Referring to Jambūdvīpa, the *Bhāgavatam* states in verse 5.17.11:

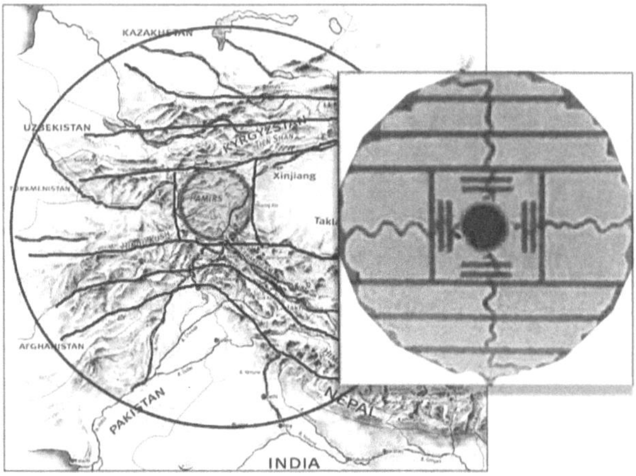

Figure 13: Jambūdvīpa as a local map of greater India. (Image used with permission from the Radha Mohan Das - Vedic Science and History Youtube Channel)

> Among the nine *varṣas*, the tract of land known as Bhārata-varṣa is understood to be the field of fruitive activities. Learned scholars and saintly persons declare the other eight *varṣas* to be meant for very highly elevated pious persons. After returning from the heavenly worlds, they enjoy the remaining results of their pious activities in these eight earthly *varṣas*.

b. Contemporary models drawn from Vaishnava culture

During the summer of 1977, Prabhupāda requested his followers to seek out scholars of Purāṇic cosmology who could help develop the proposed Temple of the Vedic Planetarium which he envisioned housing "a huge, detailed model of the universe as described in the text of the Fifth Canto of the *Śrīmad-Bhāgavatam*."[4] One Purāṇic model that offered utility was found in the 19th-century illustrations of Śrī Tiruveṅkaṭa Rāmānuja Jīyra Swami, a Vaishnava in the line of Rāmānujācārya. His presentation drew upon constructs similar to Tycho Brahe's 16th-century model and descriptions found in the *Sūrya-siddhānta*, only in Tiruveṅkaṭa's version, earth-centered perspectives of the solar system align with the Bhū-maṇḍala.[5] Danavir Goswami discusses this model in *Vedic Cosmos - The Mystery of the Workings of the Universe* (2007), while subsequently developing a number of projects that have inspired further research.

For example, Pavaneshwar's (Santhoju Pandu, MS Engineering) recent book, *Bhāgavata Cosmology: Vedic Alternative to Modern Cosmology* (2018), expands on these works by supplementing Tiruveṅkaṭa's diagrams with descriptions of the Purāṇic cosmos found in the 16th-century treatise, *Bhūgola Varṇanam* by Śrīla Vādirāja Tīrtha Swami, a Vaishnava in the line of Madhavācārya. A close analysis of the text reveals a number of parallels with contemporary astronomical calculations. In this model, Bhū-maṇḍala is depicted as a solid plane with a depth of 800,000 miles (~1.3 million km) in the center, gradually tapering as it extends toward the circular Lokāloka Mountain range, then continuing to a radial distance of 1 billion miles (~1.6 billion km). By combining this account with *jyotiṣa* calculations such as those described in the *Sūrya-siddhānta*, along with recent mathematical analysis by Punit Bhalla, Pavaneshwar demonstrates predictive values located in cosmological texts identified with the Vedic tradition that illustrate observable motion of the sun, moon, planets, and stars.[6]

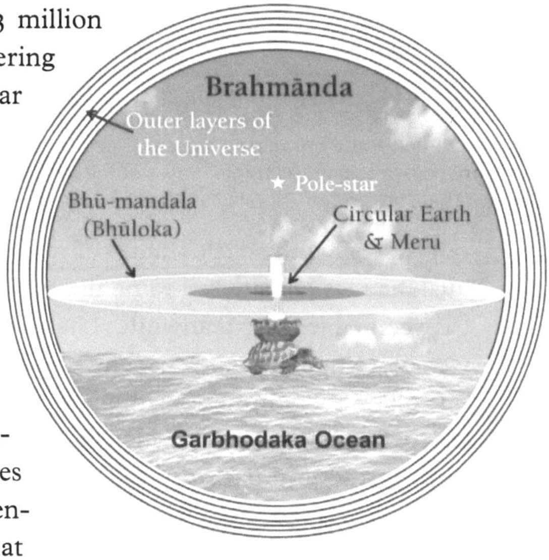

Figure 14: A vertical cross-section of the Brahmāṇḍa, illustration from *Bhāgavata Cosmology* (2019) by Santhoju Pandu.

Prem Gaurāṅga (Punit Bhalla) also drew upon similar models while exploring earth-sun parameters that correspond with observational astronomical phenomena, such as perihelion and aphelion distances, orbit eccentricity, and polar declination. In his approach, a model of Bhū-maṇḍala closely aligns with the equatorial plane while maintaining a constant 3.5° angle. Prem Gaurāṅga draws his data almost entirely from the *Śrīmad-Bhāgavatam* in his paper, "Mathematical Model of the Cosmos based on the *Bhāgavat Purāṇa*," presented at the 2019 Bhaktivedanta Institute Cosmography Workshop held in Gainesville, Florida.

c. Philosophical analyses from the eastern tradition

Figure 15: Punit Balla compares the sun's chariot to a traditional oil-pressing machine. Water color by John Gantz, "The Checkoo or Oil Mill," 1828.

Ashish Dalela (Ṛṣirāja das, MS Chemistry) adopts a semantic approach to the Purāṇic worldview in his recent book, *Mystic Universe: An Introduction to Vedic Cosmology* (2016). Rather than explore the Fifth Canto in terms of mathematical astronomy as a physical account of space, he analyzes geographic language used throughout the text with subtle frames of reference. For example, definition for "distance" is considered to be the time it takes to transfer information between objects. Meanwhile, "height" becomes successive stages of *guṇa* development, and "oceans" define the level of disconnect between one place and another. Through this lense, Bhū-maṇḍala is viewed as a vast planetary system in which "all planets are spheres from the sensual perspective, and flat from the mental viewpoint."

As described by Dalela, a "deep understanding of the theory" embedded within the *Purāṇas* becomes "essential if the model [is] to be understood." He proposes that differences arise since "space and time in Vedic cosmology" are considered "hierarchical, closed, and cyclic," while modern science and mathematical astronomy consider them "flat, open, and linear." Dalela concludes that issues involving "meaning, mind, and consciousness" intrinsic to the *Purāṇas* "entail a different view." As a result, considering the cosmos as either "with [or] without mind," can lead to two distinct approaches.

Drawing from Dalela's research, Vrindavana Priya (Venkatesh Chembrolu), based in California's Silicon Valley, has similarly embraced a consciousness-centered analysis embedded within Purāṇic and Siddhāntic models for developing a series of public lectures, seminars, and virtual presentations utilized for ISKCON outreach.

d. Proposals for a Chandelier Model

In 2010, the Temple of the Vedic Planetarium (TOVP) considered the merits of two proposals for designing a Purāṇic model of celestial motion within its central dome. Antardwīp dāsa's (Adam Carroll) paper was selected as a design proposal for a model featuring a combination of Puranic cosmological descriptions with mathematical analyses from the *Sūrya-siddhānta* to explain observational phenomena such as night and day, eclipses, and seasonal changes in relation to Fifth Canto accounts. In this presentation, the aspects of the Bhū-maṇḍala plane that appear inaccessible to standard human sense perception, can be experienced through higher order consciousness.

Figure 16: Planetarium model under consideration for the Temple of the Vedic Planetarium.

Caraṇa Reṇu (Rhiannon Dauster, PhD Astrophysics), likewise offered a chandelier proposal that garnered merit. Her paper described a model where Bhū-maṇḍala aligns with the equatorial plane. Aspects of this approach had previously been analyzed in *Vedic Cosmography and Astronomy*, and more recently in Prem Gaurāṅga's paper, which also explores a correspondence between mathematical astronomy and Bhū-maṇḍala as a plane closely aligned with the celestial equator.

e. Perspectives on the Bhū-maṇḍala plane

The Bhaktivedanta Institute for Higher Studies (BIHS) research team has been exploring a number of initiatives involving Purāṇic cosmology.

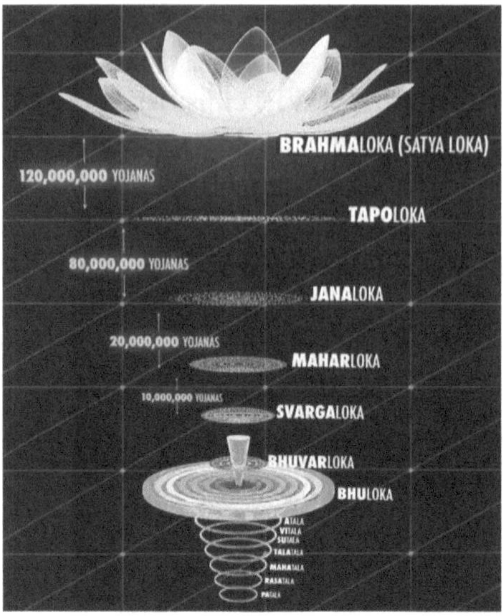

Figure 17: Visualization of the Brahmāṇḍa's vertical dimension as described in the 5th Canto of *Śrīmad-Bhāgavatam*.

Murli Gopal (Mauricio Garrido, PhD Physics) is taking a fresh approach examining mathematical relationships drawn from the perspective of Bhū-maṇḍala as the horizontal plane, as experienced by an observer. This particular vantage point may help provide correspondence between a variety of models of the Vedic cosmos. Dvija Govinda (Vasyl Semenov, PhD Physics) has concentrated his research on the vertical dimension, an integral and particularly intriguing aspect of Fifth Canto cosmography. By mathematically treating the *yojana* as a sliding scale applied to the "heights above Bhū-maṇḍala," as well as exploring space-time dilation constructs, he is finding intriguing parallels with contemporary astronomical calculations.

Vaishnav (Vaibhav Tiwari, PhD Astrophysics) and Doug Watson (PhD Physics), also on the BIHS team, are studying broader cosmological questions involving topics such as dark matter and the *pravāha* wind, the origin of the solar system vis-à-vis Purāṇic descriptions of Priyavrata's chariot wheels, and recent scientific theories involving cyclic scenarios for universal creation and destruction. Another PhD in Astronomy, Giridhāri (Tiziano Valentinuzzi), has been investigating information embedded within the Fifth Canto potentially referencing heliocentric perspectives while drawing upon research published by Roberto Caimmi, the physicist and astronomer (retired) from Padua University.[7] As for public outreach, Rādhā Mohan (Richard Cole), who also serves as Communications Officer at the Bhaktivedanta Manor, has for a number of years produced videos for his "Vedic Science" YouTube channel. He now includes BIHS research material that offers a general audience ready access to the cultural legacy of Purāṇic cosmology.

The recent work by an independent scholar based in Australia, Sulabha (Steven Doyle, pen name Soolaba), might initially appear to access flat earth constructs within a contemporary Western context. But closer analysis reveals a focus on physical representations and philosophical arguments rooted within Purāṇic and *jyotiṣa* texts. For example, in his introduction to a forthcoming volume, *The Purāṇic Geography of Bhārata-varṣa*, Sulabha writes that his work examines how "our Earth realm has very peculiar qualities." When considering the complexity inherent to modeling Bhū-maṇḍala, Sulabha proposes it "indeed" appears as "a vast immobile flat plane, yet simultaneously, has mathematical features that resemble a sphere in space." He argues that this information is found not only within Purāṇic literature, but also within observable data drawn from both *jyotiṣa* mathematical astronomy and contemporary science.

As a matter of cross-disciplinary interest, Western scholars have also commented upon a variety of ancient accounts for an expansive plane depicting an earthly realm, that compares to the plane of earth's orbit around the Sun according to mathematical astronomy. The noted MIT historian and philosopher of science, Giorgio de Santillana, along with his collaborator, Hertha von Dechend, presented a provocative analysis in their book, *Hamlet's Mill: An Essay Investigating the Origins of Human Knowledge and Its Transmission Through Myth* (1969). Though ancient civilizations are generally thought to have considered the earthly realm as a flat disk, these two authors argued that a number of apparently mythological accounts were not necessarily fictitious, but frequently represented accurate descriptions of celestial cosmology. While examining scores of sacred texts and traditional "myths" describing a broad range of ancient traditions through an astronomical lens, the authors suggest that if notions of a "cosmic earth-disk" or "'flat earth' do not correspond" to the perspectives of "flat-earth fanatics," so too "the name of 'true earth' (or of 'the inhabited world') did not in any way denote our physical geoid." The authors proposed that the expansive celestial plane described by these cultures rather corresponds to "the band of the zodiac, two dozen degrees right and left of the ecliptic."[8] Thus, they argue, so-called mythological stories may actually pertain to astronomical events involving an "ideal plane going through the four points of the year, the equinoxes and the solstices."[9]

Thompson commented in his book, *Mysteries of the Sacred Universe*, on the work of de Santillana and von Dechend: "It is noteworthy that in a cross-cultural study of myths from around the world, these two scholars found evidence for "an archaic world culture ... in which the earth was seen as the celestial plane of the ecliptic – the realm of the gods and the planets." Thompson suggested that when considering this perspective, "It may be that the concept of the naive flat earth arose when people began to take this cosmic earth-disk literally," and then essentially "identified it with the ground on which they stood."[10]

In other words, there may well be tangible historical evidence for concepts of an expansive celestial plane identified within ancient cultures from around the world that offers parallels to both Purāṇic descriptions of Bhū-maṇḍala, as well as mathematically quantifiable data found within *jyotiṣa* and more recent forms of observable astronomy.

CHAPTER 10

Empirical Evidence and Natural Science Within a Western Context

a. Introduction: Historical background of a long discussion within the West

In recent times, flat earth advocates have promoted a variety of perspectives on YouTube and elsewhere. This section will examine a number of related concerns expressed within popular debates that involve the natural sciences. While by no means a comprehensive account, it will offer a useful introductory overview.

Of note, while modern flat earth advocates consider the earth as a plane rather than a sphere, such concerns are significantly different from discussions held within the ancient tradition of the Indian subcontinent. There, Purāṇic accounts of an expansive Bhū-maṇḍala disk that stretches billions of miles invited attention with reference to the Sanskrit mathematical astronomy of the *jyotiṣa* tradition, which offers calculations of the earth as a globe.

In contrast, advocates for a contemporary flat earth model depict maps centered on the North Pole surrounded by continents and oceans. The total circumference of this model measures in the tens of thousands of miles. It also includes an impregnable wall of ice rising along the perimeter representing Antarctica, an image otherwise reminiscent of a world map featured on the United Nations flag. These arguments are identified with the work of the 19th-century Englishman, Samuel Rowbotham (1816–1884). Rowbotham produced a variety of tracts that attracted followers, primarily in Britain and America. Advocates such as

William Carpenter (1830–1896) and Lady Elizabeth Blount (1850–1935), promoted this movement based on Rowbotham's 221-page book, *Zetetic Astronomy: Earth Not A Globe!* initially published in 1865.[1] In the mid-20th century, a successor organization, The International Flat Earth Research Society, took up the mantle. Its founder, Samuel Shenton (1903–1971), along with his successor, Charles Johnson (1924–2001), considered flat earth arguments compatible with Biblical accounts, as had a number of earlier proponents.[2] Suffice it to say, accounts of this model have not been drawn from the Purāṇic tradition.

Within the long history of Western discourse, perceptions of a spherical earth were well known within the ancient world, and subsequently, to medieval scholars. Aristotle (384–322 BCE), considered Plato's foremost student, has been particularly influential within the Western tradition. For example, Christianized versions of his natural philosophy were well established within the European university system that initially arose from Roman Catholic church schools. Aristotle's reputation was held in such high esteem that during the Middle Ages he was known as "The Philosopher" amongst European scholars, while among Muslim scholars, "The First Teacher." Aristotle's views on physical science remained dominant until the early modern period, when the efficacy of classical mechanics, identified with Newton, became recognized as offering superior predictive methodologies.

As for Aristotle's arguments concerning the spherical shape of the earth, a number of them involved fairly ordinary events such as examining the appearance of a ship rising over the horizon, the crescent shape of a lunar eclipse, and the shifting position of stars at different geographic locations.[3] In other words, these cases involved evidence accessible to ordinary perception, and they predate the rise of modern science identified with European origin by thousands of years.

As an aside illustrating cross-cultural influences within the ancient world, one of Aristotle's leading students, Aristoxenus (fl. 335 BCE), is recorded as stating that Aristotle's direct predecessor, Socrates (c. 470–399 BCE), was known for his discourse with Indian *sādhus* that subsequently influenced Plato, Aristotle's teacher. In an authoritative account compiled by Bishop Eusebius of Caesarea (c. 260–339 CE)[4] – a leading scholar of his generation – Aristocles of Messene (fl. 1st century CE) is quoted as stating that while Plato made distinctions amongst categories of knowledge, he also "perceived that the science of things divine and human was one and the same," and further "maintained we could not take a clear view of human affairs, unless the divine were previously discerned." Aristocles proposed that Plato surely had been influenced by Socrates, about whom Aristotle's student (Aristoxenus) had previously written:

> this argument comes from the Indians: for a certain man of that nation fell in with Socrates at Athens, and presently asked him, what he was doing in philosophy: and when he said, that he was studying human life, the Indian laughed at him, and said that no one could comprehend things human, if he were ignorant of things divine.[5]

As for the physical science of the globe, there were numerous Greek natural philosophers offering analyses, with some calculating the earth's circumference close to modern measurements. For instance, Eratosthenes, based at the famed Library of Alexandria, calculated the size of the earth within a few hundred miles of contemporary figures. He is also considered to have accurately estimated the tilt of the earth's axis at around 200 BCE.

In other words, such concerns were well known amongst natural philosophers of the ancient world. Greeks were so familiar with the concept of a spherical planet, that Megasthenes, a Greek official at the court of Chandragupta and his brahminical advisor Cāṇakya Paṇḍita,[6] is credited with stating in his accounts of India:

> On many points their opinions coincide with those of the Greeks, for like them, they say that the world ... is in shape spherical, and that the Deity who made it, and who governs it, is diffused through all its parts.

Megasthenes is further described as noting, "Concerning generation, the nature of the soul, and many other subjects, they [the residents of India] express views much like those maintained by the Greeks."[7] The take-away point being that an appreciation of the spherical nature of the earth was considered a favorable East-West cultural comparison, dating back as early as 300 BCE. Accounts such as these suggest that an understanding of the earth as a globe is not a recent innovation.

Of additional historiographical interest, at the time of Christopher Columbus (1451–1506), few in Europe held genuine concern that his famed expedition might fall off the earth. After all, medieval scholars were well-versed in works such as those of Dante Alighieri (c. 1265–1321), the Roman Catholic author described as "the greatest of Italian poets" and whose narrative, the *Divine Comedy*, has been considered "the finest of all Christian allegories."[8] Woodcuts adorning Dante's work clearly illustrate the spherical nature of the earth.[9]

Figure 18: "Vano Dello Infero [The Chamber of Hell]," by Antonio Manetti (1423–1497). Manetti conceptualized the cosmographical dimensions Dante described in his narrative poem, the *Divine Comedy*, written by Dante c. 1321. This woodcut print illustrating the world as a sphere "is one of seven maps the appeared in the first illustrated version of Manetti's work, published c. 1506." The Cornell University Library Digital Collections offers a copy of the image along with historical information at the following address: https://digital.library.cornell.edu/catalog/ss:3293742.

For example, in the first section of the *Divine Comedy* known as *L'Inferno*, Dante described the weight of gravitational force emanating in all directions from the center (Canto 34, lines 106–115). Dante (in the role as the narrative's central character) and his guide Virgil, after viewing Lucifer imprisoned at the earth's core, subsequently returned to the surface on the opposite side of the world. In the second section, *Purgatorio* (Canto 1, lines 22–27), Dante noted the different view of the stars in the Southern and Northern Hemispheres, an expected observation for those living on a spherical planet. In *Purgatorio* (Canto 27, lines 1–6), Dante additionally wrote of a general sense of time zones when describing that as morning dawns in Jerusalem, it is high noon on the River Ganges. Meanwhile, in Spain, it is midnight. Or as Dante explained it in the early 1300s to his Medieval Christendom audience:

As when his [the sun's] *earliest shaft of light assails*
The city [of Jerusalem] *where his Maker shed His blood,*
When Ebro [in Spain] *lies beneath the lifted Scales*
[i.e., midnight]
And noontide scorches down on Ganges' flood,
So rode the sun; thus day was nightward winging
When there before us God's glad angel stood.[10]

Another influential medieval author, Johannes de Sacrobosco (c. 1195–c. 1256), has been identified as an Augustinian monk who taught at the University of Paris and wrote a number of textbooks widely used during the Middle Ages through Early Modern period. Among them was *Algorismus*, which offered an introduction to Hindu-Arabic numerals that described the superiority of using them over the old Roman system. Sacrobosco's *De sphaera mundi* (*On the Sphere of the World*), published c. 1230, was likewise required reading in European universities for nearly four centuries.[11] Its first chapter describes the earth as follows:

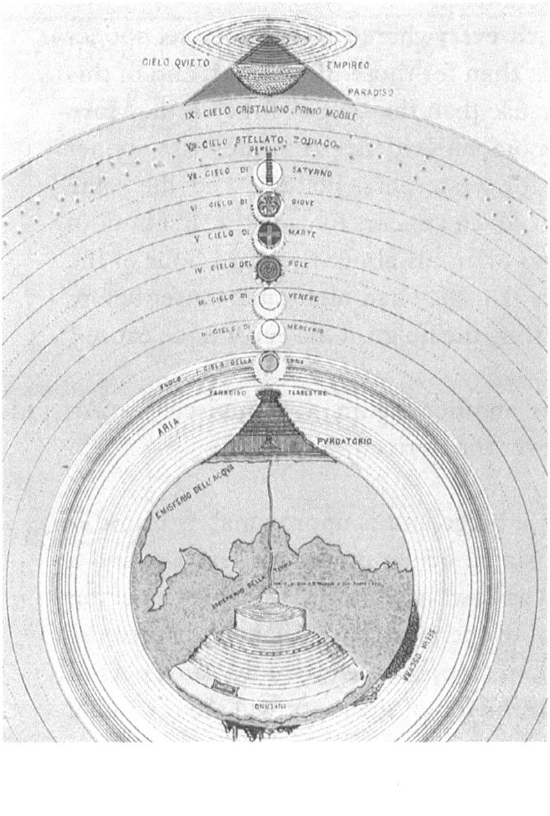

Figure 19: "Figura Universale Della Divina Commedia [Overview of the Divine Comedy]," by Michelangelo Caetani (1872). Caetani is considered a well-regarded Dante scholar, albeit active centuries after Dante initially published. The Cornell University Library Digital Collections offers a copy of this image along with information on its history at the following address: https://digital.library.cornell.edu/catalog/ss:3293735.

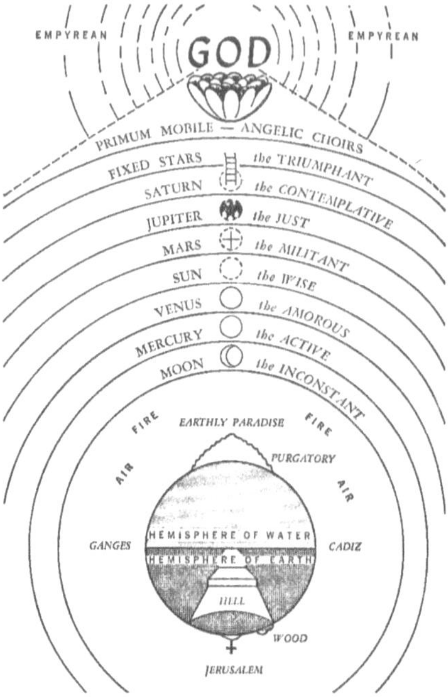

Figure 20: "Dante's Cosmos." Note: In this image, Spain is identified with Cádiz, whereas in the narrative poem Spain is identified with Ebro. The Ganges is located on the left side of the globe opposite Cádiz, with Jerusalem located at the bottom. This image was published by the respected Dante scholar, Thomas Goddard Bergin, in his book, *Dante* (New York: The Orion Press, 1965), p. 240. During his tenure at Yale University, Bergin held the academic position of Sterling Professor of Romance Languages. Bergin is also note for being the first modern poet whose words have been launched to space on a satellite in orbit of the earth.

> That the earth, too, is round is shown thus. The signs and stars do not rise and set the same for all men everywhere but rise and set sooner for those in the east than for those in the west; and of this there is no other cause than the bulge of the earth. Moreover, celestial phenomena evidence that they rise sooner for orientals than for westerners. For one and the same eclipse of the moon which appears to us in the first hour of the night appears to orientals about the third hour of the night, which proves that they had night and sunset before we did, of which setting the bulge of the earth is the cause.[12]

Sacrobosco would utilize another argument reminiscent of Aristotle involving a ship sailing over the horizon:[13]

> That the water has a bulge and is approximately round is shown thus: Let a signal be set up on the seacoast and a ship leave port and sail away so far that the eye of a person standing at the foot of the mast can no longer discern the signal. Yet if the ship is stopped, the eye of the same person, if he has climbed to the top of the mast, will see the signal clearly.[14]

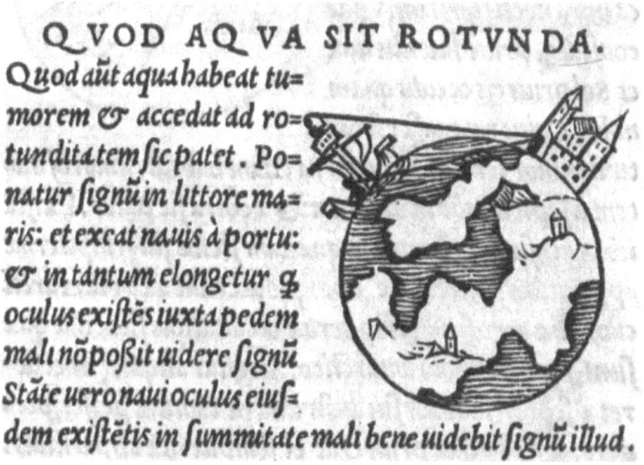
QVOD AQVA SIT ROTVNDA.
Quod aũt aqua habeat tu=
morem & accedat ad ro=
tunditatem sic patet. Po=
natur signũ in littore ma=
ris: et exeat nauis à portu:
& in tantum elongetur q̃
oculus existẽs iuxta pedem
mali nõ possit uidere signũ
Stãte uero naui oculus eius=
dem existẽtis in summitate mali bene uidebit signũ illud.

Figure 21: "Quod Aqua Sit Rotunda [That the Water is Round]," excerpt from Johannes de Sacrobosco, *De sphaera mundi* (c. 1230). This image was published in an edition edited by Petrus Apianus, Ingolstadt, Bavaria, 1526. Sacrobosco writes that while an observer on the ship's deck can no longer see a signal shining from the seacoast, another observer situated at the top of the ship's mast still can, suggesting evidence for the spherical nature of the earth. The Munich DigitiZation Center (MDZ) of the Bavarian State Library (BSB) offers a download of the Latin manuscript at the following address: https://mdz-nbn-resolving.de/details:bsb00032844.

Centuries later while considering the relevance of Columbus' expedition, the advisors to the King and Queen of Spain were hardly concerned about the shape of the earth and would rather focus on practical matters such as if the ships could carry sufficient provisions for an open ocean voyage to India and China. Though his Spanish patrons considered the earth a globe, they were not entirely sure of its size, or if the uncharted ocean currents and trade winds would prove favorable. In other words, except for a question of its relative girth, there was nothing new in 1492 (coincidently the time of Caitanya Mahāprabhu) for Columbus to prove concerning a perception of the earth as a globe.[15] Though Columbus would not discover a passage to India or China (to the disappointment of his European patrons), his attempt to reach the East by sailing west across the ocean around the globe did espy something unwittingly completely different, the Americas.

Suffice it to say, complexity abounds in historical accounts East and West. It is useful to be aware, as contemporary arguments can sometimes reference a simplistic imagined past while attempting an authoritative posture. The balance of this chapter will examine evidence drawn from the natural sciences as discussed in more recent debates.

b. Satellite images of a spherical earth

With the advent of space travel in the 1960s, numerous images are now available offering a comprehensive photograph of the earth planet. Among the most iconic is "The Blue Marble," photographed in one frame at a distance of about 18,000 miles (~29,000 km) during the December 1972 Apollo 17 mission. Numerous photographs from this period subsequently became emblematic for the then-emerging environmental movement. This section will consider an array of earth images in light of concerns challenging authenticity.

Various government space agencies admit to producing both composite images of the earth generally taken from low orbit satellites, along with one frame imagery of the entire globe from satellites at a greater distance. The very first full-disk Earth image is credited to the Soviet Union's Molniya 1-3 satellite, launched May 1966 from the Baikonur Cosmodrome (the world's first spaceport) located in Southern Kazakhstan, then part of the USSR.[16] The Molniya satellite was "the first of a series to carry a television," and from "its high apogees over the northern hemisphere" it "transmitted detailed cloudcover pictures of the entire disc of the earth."[17] The same Soviet spaceport had previously launched Sputnik 1, the world's first artificial satellite sent into orbit in October 1957 for three weeks, an event that became an international sensation. The Soviet program (Космическая программа СССР) also included the 1959 Luna 3 satellite, credited as the first to send back to earth never-before-seen images of the far side of the moon. Another achievement was the world's first human space flight on Vostok 1, carrying Russian cosmonaut Yuri Gagarin into Earth's orbit in April 1961.

A few months after the Soviet's Molniya 1-3 mission, NASA's Lunar Orbiter 1 earned credit for taking the first Earth images from the vicinity of the moon in August 1966. In December of that year, NASA's ATS-1 (Applied Technology Satellite) communications and weather satellite sent back an image of the earth globe with the moon in the background. The following year, ATS-3 produced the first full-disk color photograph from just over 21,000 miles (~34,000 km) above the earth's surface. A composite digital mosaic of the entire globe produced from the satellite's images subsequently became featured on the cover of the first edition of the *Whole Earth Catalog* (1968), the iconic journal of the American countercultural movement. It appears that NASA had not intended to release these images, and they did so only after Stuart Brant, the famed editor of the *Whole Earth Catalog*, successfully lobbied for their public viewing.

Figure 22: A composite digital mosaic image of the entire globe featured on the cover of the first edition of the *Whole Earth Catalog* (1968).

Another famous image, "Earthrise," was taken in one shot by William Anders using a modified Hasselblad 500 EL camera on Christmas Eve 1968 during the Apollo 8 mission. It likewise is considered one of the most influential environmental photographs ever taken. The United States Postal Service issued a stamp in 1969 with an image of the Apollo 8 Earthrise photograph. It included the words, "In the beginning God..." created the heavens and the earth, the opening Biblical stanza from the Old Testament.

Already mentioned is "The Blue Marble," the one-shot full disk photograph featuring a fully illuminated globe offering a clear view of Africa and Antarctica. Portions were under a cloud cover that included remnants of the 1973 Tamil Nadu cyclone. This image of the earth globe, widely released on the front page of newspapers on Christmas Day, is considered among the most distributed photographs ever taken and the most requested from the NASA archives. Though credited to the entire crew, consensus leans towards it having been taken by astronaut Harrison Schmitt using a 70-millimeter Hasselblad camera with an 80-millimeter Zeiss lens.[18] This image has also been featured on the Earth Day flag representing environmental awareness.

Figure 23: United States Postal Service First Class stamp (1969) featuring an image of the Apollo 8 "Earthrise" photograph taken in one shot by astronaut William Anders.

As noted, a number of satellite images are digital mosaics, though as also noted, not

Figure 24: An image of "The Blue Marble" photograph featured on an Earth Day flag.

all of them are, such as "Earthrise" and "The Blue Marble." More recently, the 2015 launch of the Deep Space Climate Observatory (DSCOVR) began regularly beaming one-shot full-disk images utilizing an Earth Polychromatic Imaging Camera (EPIC) from distances approaching one million miles. Though managed by the National Oceanic and Atmospheric Administration (NOAA) as a weather, climate, and earth observation satellite, NASA processes its Earth sensor data and provides a variety of high resolution photographs on its website dedicated to the project.[19] Meanwhile, the NOAA also operates the Geostationary Operational Environmental Satellites (GOES) program for weather forecasting and climate research. Its GOES-16 satellite, centered over the Americas, provides a full disk image every 5 to 10 minutes.

Not all earth globe imagery is associated with NASA or the Soviet space program. Among the more recent photographs produced by government institutions from other countries, in May 2004 the Japanese Aerospace Exploration Agency's (JAXA) robotic spacecraft Hayabusa, which while designed to study a near-earth asteroid, also took full-disk earth images from about 185,000 miles (~300,000 km). JAXA's Geostationary Meteorological Satellite "Himawari-8," launched in 2014, continues to capture geostationary imagery in real time from approximately 22,000 miles (~35,400 km) as part of its climate observation mission.

In summation, while low-orbit satellites have produced full-disk images utilizing digital mosaic and variable scanning technology, there have also been a variety of deep space missions credited with taking one-shot full-disk photographs, with a number of these images taken by human beings utilizing traditional film technology. Either way, whether taken in one frame or stitched together from a number of images, the form of the earth credited to these satellites consistently appears as a globe.

c. Curvature: Views from weather balloons, amateur rockets, and airplanes

Some argue that the main evidence for curvature of the earth comes from modern satellite imagery. Apart from photographs attributed to massive government agencies, there are also numerous videos available on YouTube produced by a variety of modest-sized professional and amateur research projects utilizing high-altitude weather balloons and small-scale rocket technology, that offer visual recordings of the curvature of the earth planet.[20]

Historically, the first direct detection of the earth's curvature occurred May 1931, as reported by the Swiss physicist Auguste Piccard and his assistant Paul Kipfer while flying over the Bavarian Alps in a hydrogen-filled balloon. Theirs was also the first manned flight to reach the stratosphere at nearly 52,000 feet (~16,800 m) utilizing a pressurized gondola.

The first photograph of the curvature is credited to the Explorer II high-altitude balloon project, launched November 1935 at the "Stratosphere Bowl" near Rapid City, South Dakota. Management of the project was coordinated by the National Geographic Society and the United States Army Air Corps (USAAC). The Explorer II gondola, presently on display at the Smithsonian National Air and Space Museum in Washington D.C., carried USAAC Captains Albert W. Stevens and Orbil A. Anderson well into the stratosphere at over 72,000 feet (~22,000 m), a record for human flight held until the 1950s. The *National Geographic* magazine's May 1935 edition subsequently included a 16.5 by 24 inch special supplement fold-out poster featuring a mission photograph of the earth's curvature, along with another of the many articles it had been publishing about the project.[21]

The Smithsonian Museum website describes the following about the program:

> One of the flight's most dramatic successes was the aerial photograph: the first photographs showing the division between the troposphere and the stratosphere and the actual curvature of the earth from the record altitude. Captain Steven's cameras captured stunning photographs of

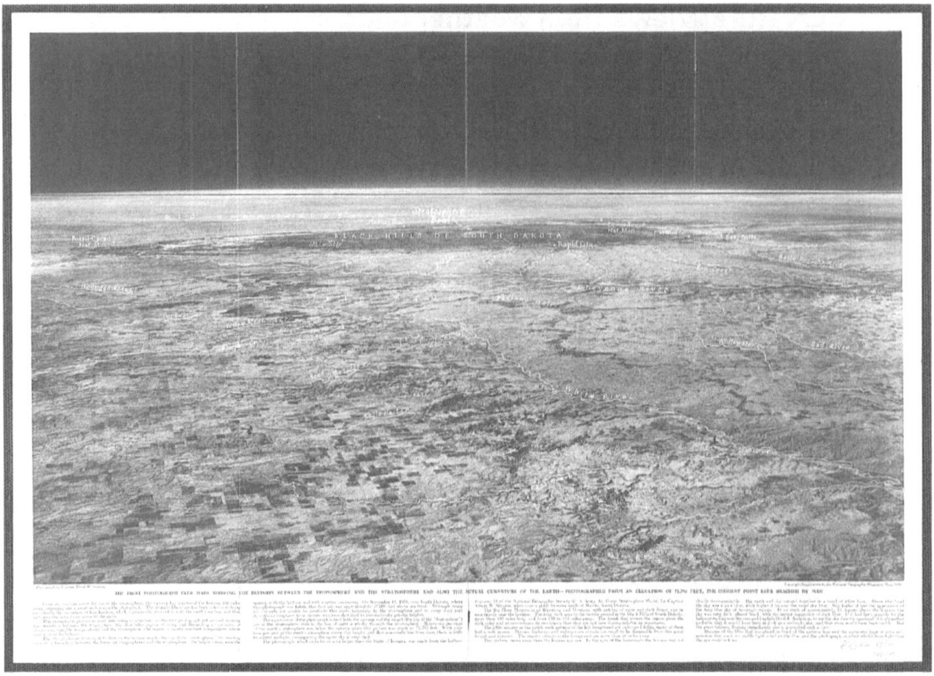

Figure 25: Image of the fold-out poster offered as a "Supplement to the National Geographic Magazine, May, 1936" with the title: "The first photograph ever made showing ... the actual curvature of the earth–photographed from an elevation of 72,395 feet, the highest point ever reached by man." This image was taken "over a point 35 miles south of Murdo, South Dakota." The text states that a "straight black line has been ruled in to bring out clearly the curvature."

> South Dakota and surrounding states, and demonstrated the potential of high-altitude, long-range reconnaissance from manned balloons.[22]

Reports described as many as 20,000 viewers at the launch site on the day of the event, which was also broadcast on live radio by the National Broadcast Company (NBC). The gondola held a small transmitter, and audiences in America and Europe could hear real time accounts of the aeronauts speaking with engineers and others notables on the ground, including a radio announcer in London. The success of the project proved to be a national affair. After landing, the two aeronauts paid a follow-up visit to President Franklin Roosevelt at the White House. They also participated in numerous celebratory events that included the reception of scientific exploration awards.

A natural question may arise for those living in the jet age, when international travel has become a norm: At what height is curvature of the globe observable from an airplane? David K. Lynch, author of the essay "Visually discerning the curvature of the Earth," published in the professional journal *Applied Optics* (December 2008), explains that while under ideal circumstances it is theoretically possible to view at lower altitudes, "pilots and high-elevation travelers reveal that few if any could detect curvature below about 50,000 ft." But "commercial jets seldom exceed altitudes of 40,000 ft.," a situation complicated by the limited field of vision (FOV) offered by comparatively small passenger windows, whose design can affect perception. Meanwhile, travelers on the supersonic Concorde jet flying at the height of 60,000 feet (~18,300 m) "routinely marveled at the curvature of the Earth" through passenger windows the size of a human hand.[23] The panorama improved while standing in the aisle looking through a series of windows, what to speak of witnessing the expansive vista offered by the cockpit – something visitors could occasionally do prior to 2001.[24] Military and private jets flying above 50,000 feet (~15,200 m) offer similar opportunities.

Lynch writes that while he experienced "a general sense of the horizon" on a commercial flight, he admits that "actually identifying the horizon's location was difficult." Lynch attributed this to

> a very low-contrast boundary in a region of the sky where there were much higher-contrast changes. There were almost always clouds on the horizon that prevented accurate horizon location.
>
> When the horizon was clear, detecting curvature from around 35,000 ft. was relatively easy, providing that a wide, unobstructed FOV was available. With a horizontal FOV of 90° or more, the curvature was subtle but unmistakable. Under similar conditions with a FOV smaller than about 60°, the curvature was not discernible. Thus, visually detecting the curvature would seem to depend on both the actual curvature and the FOV.[25]

Lynch also writes about the "possibility of seeing it from high mountain tops," and anticipates "the altitude necessary to visually detect curvature would seem to be between 14,000 and 35,000 ft."[26] He concludes "it seems well established that the curvature of the Earth is reasonably well understood and can be measured through photographs,"[27] while also noting that "detecting the curvature of the horizon directly is a complex issue, and it is further aggravated by psychological factors."[28] As a research scientist, Lynch points out that "people hope and often expect to see it and so they do, whether they actually do or not."[29] This sort of challenge likely applies to a variety of conditions, whether curved or flat or scientific or otherwise.

As discussed in prior sections, arguments for curvature, or conversely against a simplistic perception of the earth as a plane, are not new. For thousands of years within the Vedic cultural tradition, numerous *jyotiṣa brāhmaṇa paṇḍitas* had attributed a flat appearance to the limitations of ordinary human perception able to view less than 1⁄100 of the earth's circumference while standing on the surface. As such, the expectation of a relative cognizance of flatness under ordinary circumstances had been well considered during ancient times.

d. Vanishing point: The visual disappearance of objects over the horizon

Contemporary flat earth advocates at times propose that objects on the horizon vanish due to appearing smaller and smaller. In other words, at some point they reach a vanishing point from the perspective of the observer, rather than dipping below the curvature of the horizon as would be expected if the earth were a sphere.

But this does not conform to common experience, as per traditional arguments from the Greek tradition well known within medieval European universities. For example, if observers standing on a cliff along the coast were to view a ship approaching from the distance, they first would see the tallest point, the mast, prior the body of the ship. Conversely, for a receding ship, the first thing to disappear would be the lowest point, as if dipping below the crest of a hill, or in the case of the earth, its horizon.

Thus, simple naked-eye observation, as well as more recent observations abetted by telescopes, appear to confirm the phenomenon of curvature. After all, powerful telescopes that otherwise see vast expanses of outer space, do not see endless miles of the planet's surface as would be expected while viewing a plane. Under standard conditions at sea level, the viewable horizon is limited to approximately three miles.

e. Impossible gravity: Traditional and contemporary theory

Certain contemporary flat earth advocates propose an absolute up and down to cosmic space. As a result, those on the Southern Hemisphere of a spherical planet should fall off. In response, both traditional and contemporary theories of a spherical earth frequently identified with the West propose that the central locus of gravitational force emanating from the center of the planet produces the experience of being pulled down anywhere along its outer surface. Whereas the Aristotelians of prior ages argued that the center of the earth was the center of the cosmos, in more recent times Newtonian physics posits that every object exerts some gravitational force. On a spherical planet with the central focus of gravitational force at its center, the effect would exert near equally along its surface at similar elevations.

Meanwhile, on a spherical planet, gravitational force quantified in terms of the planet's mass normally overwhelms the gravitational effect of smaller objects near its surface. In other words, an apple dropped from the hand of a child falls to the ground, whereas feathers on the ground would not rise to the underside of an apple held in the hand. The reason given is that the overwhelming mass of the earth, and thus its comparatively massive gravitational effect, would pull both the apple and the feather down toward the direction of Earth's center.

f. The orbit of the moon in relation to the earth

> "If the center of a spherical earth is the center of its gravitational force, then why would the moon maintain its orbit at an arbitrary distance?"

As per fundamental arguments drawn from classical Newtonian physics, if not for the earth pulling the moon towards its center, the natural trajectory of the moon, or any object, would be a straight line in space.

In other words, the acceleration of the moon's trajectory is balanced by the mass of the earth's gravitational pull, producing an approximately circular lunar orbit. This effect is somewhat comparable to twirling a ball on a string, with the string representing gravitational force holding the ball in a circular motion. If not for the pull of the string, the ball would assume a comparatively straight path. In the case of the moon, its acceleration is balanced by the earth's gravitational pull, producing a roughly circular orbit averaging just under 240,000 miles (~390,000 km) from the earth.

Newton's achievement involved mathematically quantifying "laws of motion" that could consistently predict the movement of earthly and celestial objects. As for its earthly effect, these principles apply equally to both cannonballs and intercontinental ballistic missiles. Of note, Newton never claimed to understand what produced gravity. He rather argued that he could reliably predict its effect relative to mass and acceleration. Newton famously feigned no hypothesis ("*hypotheses non fingo*") for the cause of gravity.

g. Lunar eclipses and the stars

Another argument identified in the West with the Greek tradition involves eclipses. Aristotle's *On the Heavens* (350 BCE) offers an explanation that "during a lunar eclipse, the earth casts a round shadow on the moon which could only be caused by the earth being a sphere."[30] This would be expected since no matter its relative position, a three-dimensional sphere casts a crescent shadow. On the other hand, a circular disk would only cast a curved edge if positioned relatively perpendicular to the illuminating object, in this case the sun, and the area shadowed, in this case the moon.

Another argument attributed to Aristotle (and his pre-Socratic predecessor, Parmenides) asserts that if the earth were a flat disk, the same set of stars would appear in the same location in every geographic region. But that is not the case, as "different heights of the stars" appear "in various parts of the world."[31] For example, the position of the pole-star above the horizon observed at the Acropolis in Athens, is different when observed at the same time and season along the River Nile at the Valley of the Kings near Thebes, well over a thousand miles to the south.

Meanwhile, in the modern era of global travel, countless star-gazers have experienced how the stars visible in the Southern Hemisphere are different from those visible in the North, and further, their motion appears opposite in direction. Such phenomena would be expected while observing the heavens from the surface of a globe.[32]

h. Concluding thoughts

This section is meant to offer a few talking points frequently raised in contemporary debates involving the shape of the planet earth. Admittedly, the weight of historical and astronomical evidence appears to play well with perceptions of the earth as a globe.

But scientific paradigms regularly shift as more information becomes available, and a more powerful explanatory hypothesis may potentially develop. Standard human perception is never assured of having access to all possible contingencies at any given moment, what to speak of explaining it definitively. Thus, scientific analyses dependent on relative human experience are bound within the limitations of conditional perception.

In other words, on these grounds, none of the arguments discussed in this chapter are intended to be considered absolutely definitive. Yet interestingly enough, a selection of the empirical evidence that challenges simplistic flat earth constructs has remained nearly the same since the time of Aristotle around 350 BCE, through to the launch of the Deep Space Climate Observatory in 2015. This suggests consistent reasoning on behalf of a spherical planet, as drawn from natural scientific observations and basic mathematical principles that have unquestionably predated the global influence of Euro-American culture by thousands of years.

Therefore, this section focuses upon a handful of prominent talking points, both traditional and modern. The idea is to familiarize readers with basic discussions on the shape of the earth frequently identified within a Western context, and to assist them to proactively engage such concerns.

Yet the question remains: What is behind the propensity of certain advocates for the Purāṇic tradition to sometimes identify the Bhū-maṇḍala plane with a geographic flat earth? Perhaps this proclivity could be triggered in part by suspicions of a perceived academic hegemony that upholds science as an infallible bastion of objective human knowledge, a posture that elicits a desire to push back. Alongside that, there may well be genuine concerns that ill-considered technological innovations identified with scientific progress could potentially inflict dire repercussions upon the planet. That in turn could engender yearnings for a simpler life identified with previous generations. It is also possible that sometimes these issues are felt more keenly within a devotional community, as a materialistic worldview often evangelized in the name of professional science can sometimes appear hand in glove with initiatives to erode human values and spiritual vision. While there are serious discussions to be considered here, accusing science of perpetuating a global hoax that the geographic earth is a sphere could well miss the mark for engendering productive resolutions.

While for followers of the Vaishnava tradition śāstric authority supersedes ordinary sense perception, claims for a geographic flat earth appear neither in accord with ordinary perception or śāstric authority as considered by Vaishnava *ācāryas*. Such ideas are hardly essential for authentically upholding the integrity of the Fifth Canto of *Śrīmad-Bhāgavatam*.

SECTION IV

Conclusion and Appendices

CHAPTER 11

Conclusion

For thousands of years the cosmological sections of *Śrīmad-Bhāgavatam*'s Fifth Canto have drawn attention from followers of the *Purāṇas*, Sanskrit mathematicians and astronomers, and more recently, Western scholars. A number of traditional Sanskrit manuscripts utilized the term *virodha-parihāra* while acknowledging disparities between the Purāṇic accounts of a massive Bhū-maṇḍala plane containing a region of human habitation identifiable with earthly experience, and the *jyotiṣa* accounts of a mathematically quantifiable earth globe familiar to contemporary audiences. This presentation has examined this situation as a heuristic device useful for not only reflecting upon issues specific to this particular tradition, but to also perhaps more broadly examine deep-seated human proclivities for analyzing their place within the grand cosmos.

Some who identify with the Purāṇic tradition may at times feel inclined to attempt to uphold its integrity by challenging a spherical account of the earth as a planet. When addressing a concern of this nature, honorable Vaishnava practitioners maintain the principle of guru, *sādhu,* and *śāstra* for evoking a mature perspective. These principles would hold a privileged position compared to empirical knowledge-acquiring methodologies subject to four recognized conditional defects described as: "subject to committing mistakes [to err is human], to becoming illusioned, to cheating and to having imperfect senses."[1] That being said, as for the question of the earth perceived as a sphere in apparent contrast to Purāṇic descriptions of a massive cosmic Bhū-maṇḍala plane: Would an authoritative Vedic analysis consider these two well-established accounts within this tradition mutually exclusive?

Ironically, advocates for flat earth perspectives are themselves subject to the same conditional shortcomings as the empirical geographers and astronomers they might wish to criticize. Possibly with similar concerns in mind, the central speaker of the *Bhāgavatam*, Śukadeva Gosvāmī, explains to Mahārāja Parīkṣit the parameters of his Fifth Canto cosmological account:

> My dear King, there is no limit to the expansion of the Supreme Personality of Godhead's material energy ... no one could possibly explain it perfectly, even in a lifetime as long as that of Brahmā. No one in the material world is perfect, and an imperfect person could not describe this material universe accurately, even after continued speculation. O King, I shall nevertheless try to explain to you the principal regions ... with their names, forms, measurements and various symptoms.[2]

Śrīla Śukadeva Goswāmī offers caution to avoid conceit while studying the *Bhāgavata Purāṇa*, and particularly for those who may wish to intellectually capture its contents while over relying upon personal experience. After all, even Śukadeva does not claim a definitive appreciation of the "limit to the expansion of the Supreme Personality of Godhead's material energy," but instead proposes he will "try to explain" a subject that "no one could possibly explain ... perfectly." Śrīla Prabhupāda wrote in his purport to this verse: "Śrīla Śukadeva Gosvāmī admitted that to give full details of this expansive material universe would be impossible, but nevertheless he wanted to give the King as much knowledge as he had received through the *paramparā* system." Prabhupāda offered as his conclusion, "The limits of the expansions of Govinda, the Supreme Personality of Godhead, cannot be estimated by anyone, even a person as perfect as Brahmā, not to speak of tiny scientists." Prabhupāda recommended that to be "satisfied with the information obtainable from Vedic sources as spoken by authorities like Śukadeva Gosvāmī" would prove sufficient.[3]

When considering that Śukadeva Gosvāmī was directly addressing Mahārāja Parīkṣit, a powerful emperor literally about to leave the world behind, along with an assembly of illustrious saints, sages, mystics, and yogis, perhaps it would be naive to expect an elementary school textbook geography lesson. While the assembly no doubt held keen interest in esoteric Purānic topics aimed toward self-realization, they could also appreciate the observable exoteric realm quantified by Sanskrit mathematical astronomy. Within a Vedic cultural context, after all, the *jyotiṣa* tradition also proved integral to all sorts of Vedic practice such as the timing of religious rituals and the casting of natal charts.[4] Those listening to Śukadeva Gosvāmī speak on Purānic cosmology knew well that his presentation, spoken within the Fifth Canto, was not that.

Section I, "Guru, *Sādhu*, and *Śāstra*," examined whether established Vedic authorities customarily disregarded an understanding of the earth as a sphere, while considering the earthly realm relative to Purānic descriptions of a vast cosmic Bhū-maṇḍala plane. Śrīla Prabhupāda, as ISKCON's Founder-*Ācārya*, repeatedly discussed both perceptions. He additionally requested that his scholarly followers soberly analyze the cosmological descriptions of the Fifth Canto for public outreach. Much like Śukadeva Goswāmī, Śrīla Prabhupāda never claimed full cognization on a topic that "cannot be estimated by anyone," while also instructing his followers to have no doubt about the authenticity of these accounts.[5]

Of note, Chapter 1 specifically explored Prabhupāda's level of comfort discussing exoteric earth-as-globe constructs, all the while upholding esoteric Purāṇic accounts of a massive cosmic Bhū-maṇḍala plane. This should not prove a surprise, as Śrīla Prabhupāda's own spiritual master, Śrīla Bhaktisiddhānta Sarasvatī Ṭhākura, had been a leading *jyotiṣa* astronomer expert at mathematically quantifying the observable realm. Meanwhile, Śrīla Bhaktisiddhānta Sarasvatī had also authored Bengali translations and commentaries on Fifth Canto cosmography, which Prabhupāda drew from almost near exclusively for his English presentation on the subject. Both *ācāryas* appreciated both worlds with intimate familiarity.

Chapter 2 offered a variety of accounts from Vaishnava *ācāryas* illustrating how traditional commentators hardly felt averse to considering the spherical nature of the planet. Chapter 3 examined specific statements from the *Śrīmad-Bhāgavatam* (a central Purāṇic text) and the *Sūrya-siddhānta* (a central *jyotiṣa* text). Descriptions from the former did not appear essentially incompatible with a spherical description of the earth, and in any case, they most certainly did not promote the simplistic flat earth model identified with contemporary discourse. In contrast, numerous analyses situated firmly within the Vedic cultural tradition drew from mathematics and logic while describing the spherical nature of the earth planet.

Section II on "Purāṇic and *Jyotiṣa* Accounts," offered an introductory overview of the Indian subcontinent's cultural legacy practiced by historically world famous Sanskrit mathematicians intimately familiar with exact mathematical astronomical accounts that considered the earth as a globe. From an ISKCON-centric perspective, most notable among these *jyotiṣa brāhmaṇas* was Bhaktisiddhānta Sarasvatī Ṭhākura, ISKCON's *param-guru*. Suffice it to say, Prabhupāda enthusiastically appreciated his life work when he stated that Śrīla Bhaktisiddhānta Sarasvatī was a "great professor of astronomy and mathematics ... who was our merciful spiritual master."[6]

To be sure, the *Śrīmad-Bhāgavatam* "described by Mahāprabhu Śrī Caitanya as the spotless *Purāṇa*,"[7] holds precedence within Gaudiya Vaishnavism for articulating the *bhakti* tradition. Some might argue that the *Bhāgavatam* should also take absolute precedence on details of observable astronomy and thus supersede the information presented in other ancient texts such as *Jyotiṣa-vedāṅga*. But neither the Vaishnava *ācāryas*, nor the *jyotiṣa brāhmaṇa paṇḍitas*, insist on such an approach. Rather, on numerous occasions they present themselves as appearing comfortable with seemingly divergent perspectives while acknowledging both Bhū-gola as a mathematically quantifiable earth globe floating in space, along with Purāṇic descriptions of a massive Bhū-maṇḍala plane expanding for billions of miles while manifesting exotic otherworldly cosmic features, as well as familiar realms of earthly experience.

For well over a millennium, practitioners who identified with the Purāṇic and *jyotiṣa* legacies have participated in a historically recorded discourse engaging with both sets of experience in a manner that suggests an intellectually essential feature of the Vedic tradition. Appreciating such concerns as contemplated, analyzed, debated, and accommodated by great minds identified with the past can perhaps help enhance present day examinations of Purāṇic cosmography and Sanskrit mathematical astronomy in a manner that likewise upholds the integrity of both.

In the years following Śrīla Prabhupāda writing to Svarūpa Dāmodara in a notable letter:

> Now our PhD's must collaborate and study the Fifth Canto to make a model for building the Vedic Planetarium ... a working model of the universe. If we can explain the passing seasons, eclipses, phases of the moon, passing of day and night, etc. then it will be very powerful propaganda.[8]

ISKCON scholars have been actively developing cosmological models drawn from accounts described in the *Bhāgavata Purāṇa*. Chapter 9 briefly examines a selection of contemporary presentations that consider traditional Purāṇic and *jyotiṣa* accounts along with more recent Western models. These works engage a variety of analyses that accommodate both Purāṇic descriptions of the Bhū-maṇḍala plane along with empirical perceptions of a spherical earth planet.[9]

The final chapter explores perspectives drawn from natural science frequently raised by contemporary flat earth advocates. Of note, there are a number of references describing a spherical planet that can be traced in the West to ancient Greece, indicating that much like in the East, descriptions of the earth as a globe are hardly new. Although scientific paradigms may shift over time, a number of arguments have remained consistent through the ages that draw upon empirical evidence considering the earth as a spherical planet.

It is impossible to know with certainty what future generations may entertain involving perceptions of the natural world. Hence this essay does not attempt to definitively argue all that could potentially be, for it always remains possible that the full mystery may forever prove elusive, and in particular for natural philosophers whose experience is limited by the restrictions of conditional sense perception. As noted, Śukadeva Goswāmī plainly states that even Lord Brahmā, a divine personality traditionally upheld as possessing the fullest faculties as chief cosmic engineer, still lacks perfect knowledge about this topic.[10] It has also been stated that the "expansions of Govinda cannot be estimated by anyone,"[11] and even more so, that the Supreme Personality of Godhead's quality of omniscience to fully appreciate Himself is in an eternal competition with His ever-expanding creative potencies.

Gaining a glimpse of how these issues have been considered within a traditional cultural setting identifiable with the Sanskrit term *virodha-parihāra*, may well prove useful to contemporary discussions by offering contextual intellectual perspective. After all, would it really be safe to assume that modern scientists are boundless leaps ahead of past natural philosophers in a quest for certainty, especially when considering the seemingly unlimited potential of all that can possibly be? For example, views of the cosmos appraised by 20th-century astronomers have arguably shifted far more dramatically from the views held by natural philosophers during the 17th-century Scientific Revolution, then the prior shift from Aristotelian geocentric perspectives to Newtonian heliocentric analyses.[12] With such thoughts in mind, who can definitely predict what the future may hold?

When considering the seemingly endless complexity inherent to any investigation of the natural world, perhaps Jñānarāja, the 16th-century *jyotiṣa brāhmaṇa* who lived during the time of Caitanya Mahāprabhu, was onto something when he considered that in the end, the "brilliance" of Śri Kṛṣṇa "drowns out ... the light of the sun, moon, and stars."[13] Bhaktivedanta Swami Prabhupāda similarly wrote, "Any glorious or beautiful existence should be understood to be but a fragmental manifestation of Kṛṣṇa's opulence, whether it be in the spiritual or material world."[14] For as described by Śrī Kṛṣṇa in *Bhagavad-gītā* Chapter 10, verse 41: "Know that all opulent, beautiful and glorious creations spring from but a spark of My splendor."

"The frog in the well" logic illustrates that a frog residing in the atmosphere and boundary of a well cannot imagine the length and breadth of the gigantic ocean. Such a frog, when informed of the gigantic length and breadth of the ocean, first of all does not believe that there is such an ocean, and if someone assures him that factually there is such a thing, the frog then begins to measure it by imagination by means of pumping its belly as far as possible, with the result that the tiny abdomen of the frog bursts and the poor frog dies without any experience of the actual ocean.

Śrīmad-Bhāgavatam 2.5.10, Purport

APPENDIX 1

Excerpts from the following essay are drawn from Bhūrijana Dāsa's *The Veil Begins to Part.*

A Lightning-Quick Tour of Canto Five

By Bhūrijana Dāsa

Section I

The philosophical thrust in Śukadeva Gosvāmī's description of the universe

After Śukadeva Gosvāmī finishes his narration of Jaḍa Bharata's instructions to King Rahūgaṇa in the Fifth Canto, the canto shifts to a different theme, with Mahārāja Parīkṣit asking Śukadeva to speak about the seven islands of Jambūdvīpa. This is because earlier he had heard Śukadeva say that these islands had been carved from the deep impressions made by King Priyavrata's chariot wheels as he followed the sun.

The king then asks a second, extremely important question: How can we see the Supreme Personality of Godhead in the feature in which He pervades the universe? The king inquires: "When the mind is fixed on the Supreme Personality of Godhead in His external feature made of the material modes of nature – the gross universal form – it is brought to the platform of pure goodness. In that transcendental position, one can understand the Supreme Personality of Godhead, Vāsudeva, who in His subtler form is self-effulgent and beyond the modes of nature. O my lord, please describe vividly how that form, which covers the entire universe, is perceived." (SB 5.16.3)

Inspired by this question, Śukadeva Gosvāmī describes both the structure of the universe and the inconceivable potency of the Supreme Personality of Godhead by which He supports and permeates the entirety of material existence. Śukadeva's almost three-hundred-verse answer takes us from chapter 16 to the end of the canto, in chapter 26.

In these eleven chapters, Śukadeva relates his divinely inspired vision of the physical universe, and along with his description the great narrator of *Śrīmad-Bhāgavatam* provides a philosophical narrative meant to carry his listeners to Kṛṣṇa, the Supreme Personality of Godhead, as He will reveal Himself in Canto Ten.

Yet on hearing Śukadeva's descriptions of the universe, modern *Bhāgavatam* readers find themselves confronted with a universe quite different from the one years of schooling have taught them to perceive.

Through Śukadeva's descriptions of the physical universe – and Śrīla Prabhupāda's powerful purports to them – the *Bhāgavatam* expands on the primary instruction Ṛṣabhadeva Mahārāja gave earlier to His sons in the fifth chapter of Canto Five: there is no happiness anywhere in the universe; whatever glimmer of happiness we can find from the upper to the lower planets cannot give us the satisfaction we crave.

So if there's no point in searching for happiness, which, as Śukadeva tells us, is unavailable, what, then, should we pursue as the goal of life? Throughout all his ensuing descriptions and his narration of various devotees' prayers, Śukadeva Gosvāmī will inspire us to seek shelter in devoted service to Kṛṣṇa regardless of the realm in which we live, for that alone produces spiritual bliss in the soul.

Bhakti, then, plus the Supreme Personality of Godhead's presence throughout His creation (*acintya-bheda* and *abheda-tattva*) – His simultaneous oneness and difference from everything that exists, which is the sublime philosophical doctrine taught by Śrī Caitanya Mahāprabhu – is the philosophical message threaded throughout the Fifth Canto and its descriptions of the universe.

One then asks: If the goal is *bhakti* and to gain an understanding of the Supreme Lord's omnipresence, should it even matter to us that the physical details Śukadeva gives are different from what modern science presents? Śrīla Prabhupāda once told a *sannyāsī* disciple, "Actually, I am not giving any more information – simply what is there." He added, frankly, "I am not an astronomer, so I may not understand all the details. Those details are given in another part of the *Vedas.*

Śukadeva Gosvāmī is giving some basic summary of the ideas of the universal operation. The stress," Śrīla Prabhupāda concluded, "is on devotion, not geography." (*The Great Transcendental Adventure,* Kūrma Dāsa, Perth, 1975, chapter 10)

The epistemological issue that underlies Śukadeva Gosvāmī's description of the physical universe

Still, it's natural to wonder about the *Bhāgavatam's* descriptions. Where does the universe come from? How does it function? How is it structured? Does anything lie beyond its unimaginably distant borders?

Śrīla Prabhupāda speaks about these questions in his purport to *Śrīmad-Bhāgavatam* 1.1.1: "It is natural that a philosophical mind wants to know about the origin of the creation. At night he sees the stars in the sky, and he naturally speculates about their inhabitants. Such inquiries are natural for man because man has a developed consciousness which is higher than that of the animals."

Śrīla Prabhupāda continues: "*Śrīmad-Bhāgavatam* gives a direct answer to such inquiries. He says that the Lord Śrī Kṛṣṇa is the origin of all creations. He is not only the creator of the universe but the destroyer as well. The manifested cosmic nature is created at a certain period by the will of the Lord. It is maintained for some time, and then it is annihilated by His will. Therefore, the supreme will is behind all cosmic activities."

But we may wish to probe further: Is the scientific viewpoint currently accepted worldwide more refined than the description the *Bhāgavatam* offers? Which view is more accurate? Is it reasonable to assume that our universe – a universe where everything that happens has a cause that can be traced to some catalyst within the universe itself – has sprung from nothing?

Modern science has shown that consciousness influences both physical reality and how we perceive it. How much, then, can our understanding of the universe ever be truly objective? How much of what we have deemed objective is tinted by the overlaying of the moods and modes that cover our consciousness? One may therefore wonder whether any scientific study of matter can ever be fully accurate.

Modern physics tends to paint for us a dark, empty, randomly generated universe, and the theories that result from this vision tend to stray from testable science into a world of ever-changing, theoretical models that are no less metaphysical than what *Śrīmad-Bhāgavatam* presents.

A number of cosmologists agree. As George Ellis writes in his article, "Does the Multiverse Really Exist," published in the August 2011 issue of *Scientific American,* "Nothing is wrong with scientifically based philosophical speculation, which is what multiverse proposals are. But we should name it for what it is."

A similar sentiment was expressed by Paul Davies in his April 12, 2003 op-ed piece, "A Brief History of the Multiverse," published in *The New York Times:*

> For a start, how is the existence of the other universes to be tested? To be sure, all cosmologists accept that there are some regions of the universe that lie beyond the reach of our telescopes, but somewhere on the slippery slope between that and the idea that there is an infinite number of universes, credibility reaches a limit. As one slips down that slope, more and more must be accepted on faith, and less and less is open to scientific verification. Extreme multiverse explanations are therefore reminiscent of theological discussions. Indeed, invoking an infinity of unseen universes to explain the unusual features of the one we do see is just as ad hoc as invoking an unseen Creator. The multiverse theory may be dressed up in scientific language, but in essence it requires the same leap of faith.

Inevitably, the attempt to ascribe truth to any description of the universe quickly sends us into questions about epistemology. His Grace Ravīndra-svarūpa Prabhu, a dear friend, recently brought this issue to my attention. Here is a paraphrase of several of his points:

- You just cannot ignore the fact that the universe as described in the Fifth Canto is neither the universe we see nor the one we experience. This is because we don't see it as *acintya-bhedābheda-tattva,* simultaneously one and different from God. God is transcendental to the world, but the world is also included in Him because Kṛṣṇa and His energies are nondifferent. Nothing is different from Kṛṣṇa, but Kṛṣṇa is different from everything. When you see in that way, you start to see the universe from another perspective. According to the *Bhāgavatam,* not only haven't we visited the moon yet, but we haven't even been to the earth.

- We have to confront the fact that we have not seen things as they really are. For example, an entire world of *devatās* exists – 330 million of them – who control our senses. That is, they control how and what we see.

- Learning to see ourselves and everything else in relationship to the Lord's universal form will help us develop our seeing and getting to know Kṛṣṇa. But the process of perception demands a purity of consciousness. Our four regulative principles are not merely moral principles but principles of cognition. Part of our process of knowing means learning to scientifically explore the universe by ridding ourselves of the modes of passion and ignorance. Then, after being situated in goodness, we move on to the state of pure goodness (*viśuddha-sattva*). In that state we can begin to see, somewhat, the way God sees.

Scientific viewpoints lean toward atheism, and we *bhaktas* do not. But like the scientists, we understand that what people experience through their senses is affected by their own consciousness, the consciousness of the observer. So what did Śukadeva Gosvāmī see? We *could* choose to accept the elegant universe as he described it because it was born of his pure consciousness. His vision leads us away from an untenable atheism, in which existence appears from nothing and has no ultimate cause or purpose, to a universe that increases our awareness of and love for God.

It's clear what position Śrīla Prabhupāda took on this choice, and he told us more than once that he accepts the accuracy of the *Bhāgavatam's* descriptions. Here is one such statement: "Modern scientific calculations are subject to one change after another, and therefore they are uncertain. We have to accept the calculations of the Vedic literature.

These Vedic calculations are steady; the astronomical calculations made long ago and recorded in the Vedic literature are correct even now. Whether the Vedic calculations or modern ones are better may remain a mystery for others, but as far as we are concerned, we accept the Vedic calculations to be correct." (SB 5.22.8, purport)

Śrīla Prabhupāda's purport to *Śrīmad-Bhāgavatam* 5.16.4 gives us a prescription: "Śrīla Śukadeva Gosvāmī admitted that to give full details of this expansive material universe would be impossible, but nevertheless he wanted to give the king as much knowledge as he had received through the *paramparā* system.... The limits of the expansions of Govinda, the Supreme Personality of Godhead, cannot be estimated by anyone, even a person as perfect as Brahmā, not to speak of tiny scientists whose senses and instruments are all imperfect and who cannot give us information of even this one universe. *We should therefore be satisfied with the information obtainable from Vedic sources as spoken by authorities like Śukadeva Gosvāmī.*"

Section II

A Brief Philosophical Overview of Canto Five

The cosmological element in *Śrīmad-Bhāgavatam* is necessary to the *Bhāgavatam's* presentation and not extraneous to it, if, for no other reason, then that cosmology is required for the *Bhāgavatam* to be established irrefutably as a *mahā-purāṇa* (see *Śrīmad-Bhāgavatam* 12.7.9–10, verse and purport). But even more importantly, *Śrīmad-Bhāgavatam's* presentation of cosmology in eleven of these Fifth Canto's twenty-six chapters (16–26) provides a background for the essential philosophical thrust Canto Five contributes to the *Bhāgavatam's* march toward Canto Ten, where Kṛṣṇa, in His beautiful, two-armed form, the *svayaṁ bhagavān,* will be presented as the summum bonum, the Absolute Truth, the highest understanding of the Supreme Personality of Godhead.

Therefore the philosophy offered in the Fifth Canto should be presented alongside the cosmology to parallel the *Bhāgavatam's* presentation in both its texts and in Śrīla Prabhupāda's purports.

Here is a lightning-quick overview of the Fifth Canto's philosophical thrust:

Śukadeva Gosvāmī dedicates eleven chapters of the Fifth Canto to Mahārāja Parīkṣit's question, asked in *Śrīmad-Bhāgavatam* 5.16.3: How can one see the effulgent Supreme Lord in the world of matter?

Śukadeva Gosvāmī begins his answer by referring to the opening cantos of *Śrīmad-Bhāgavatam,* where a spiritual aspirant is taught to view the material universe as the *virāṭ-rūpa,* or universal form, and thus learn to see the Lord everywhere.

Śukadeva then speaks about elevating one's realization to the next step: "If one reads the description of this external form of the Lord with great faith, or if one hears about it or explains it to others to propagate *bhāgavata-dharma,* or Kṛṣṇa consciousness, his faith and devotion in spiritual consciousness, Kṛṣṇa consciousness, will gradually increase. Although developing this consciousness is very difficult, by this process one can purify himself and gradually come to an awareness of the Supreme Absolute Truth." (ŚB. 5.26.38)

Then, Śukadeva Gosvāmī defines a devotee as one who aspires to attain the spiritual world as opposed to finding a safe, enjoyable place in the realm of matter. But how do those who wish to become devotees progress toward such a clearly spiritual goal? Śukadeva describes the next step in the progression: "Such a person should first control his mind by thinking of the *virāṭ-rūpa,* the gigantic universal form of the Lord, and then gradually think of the spiritual form of Kṛṣṇa [*sac-cid-ānanda-vigraha*] after hearing of both forms. Thus one's mind is fixed in *samādhi*. By devotional service one can then realize the spiritual form of the Lord, which is the destination of devotees. Thus his life becomes successful." (ŚB. 5.26.39)

Śukadeva Gosvāmī reiterates that the universe, as he described it earlier, is the Lord in His form as matter: "My dear king, I have now described for you this planet earth, other planetary systems, and their lands [*varṣas*], rivers, and mountains. I have also described the sky, the oceans, the lower planetary systems, the directions, the hellish planetary systems, and the stars. These constitute the *virāṭ-rūpa,* the gigantic material form of the Lord, on which all living entities repose. Thus I have explained the wonderful expanse of the external body of the Lord." (ŚB. 5.26.40)

Śukadeva's description of the universe is set in Canto Five after Ṛṣabhadeva Mahārāja's teachings, which occur in the fifth chapter. Ṛṣabhadeva teaches his sons that the purpose of life is to voluntarily perform austerity to attain the true fruit of a human birth. Human life, he instructs, is not meant for pursuing material pleasures, which are available even to those in the bodies of dog and hogs, but to awaken one's relationship with the Supreme Personality of Godhead and then serve Him in unlimited, unending spiritual bliss.

After hearing from his father about what should and shouldn't be done, Bharata Mahārāja, Lord Ṛṣabhadeva's foremost son, will, in his birth as Jaḍa Bharata, teach those same points to King Rahūgaṇa.

Śukadeva Gosvāmī then provides a top-to-bottom description of the universe to illustrate Ṛṣabhadeva Mahārāja's teaching that there is no true happiness to be found in any material situation on any planet anywhere. Every plan for unending happiness is spoiled by the unavoidable death of the body. From Śukadeva's description, including his description of the various hells, it's clear that death and the karmic reactions we incur for all the actions we perform in each of our bodies, stalk us relentlessly from one situation to another, from one material planet to another. No one, therefore, can be free of fear, and where there's fear, there can be no perfect happiness. So rather than presenting the universe as a facility designed to provide us with uninterrupted pleasure, Śukadeva Gosvāmī offers a far more accurate and hopeful understanding: the universe is Vāsudeva-Kṛṣṇa's energy, His *virāṭ-rūpa,* and its proper use is in His service.

Śrīla Prabhupāda shared the secret to learning to see with this understanding in a lecture he gave on Queen Kuntī's prayers in Los Angeles on April 14, 1972:

> If you love your child, if you see any garment, a small ship, shoes, or any playthings, immediately you remember your child: "Oh, this is my child's shoes. This is my child's playing things. This is his garment." So it is the custom of love.
>
> If you actually love God, Kṛṣṇa, then you can remember Him at any time, always, always. There is no difficulty.... So Kṛṣṇa is available in any circumstance—If we want to catch Him. That's all. *Teṣāṁ satata-yuktānāṁ bhajatāṁ prīti-pūrvakam.* If one is actually very serious to search out Kṛṣṇa, Kṛṣṇa is everywhere. *Aṇḍāntara-stha-paramāṇu-cayāntara-sthaṁ govindam ādi-puruṣaṁ tam ahaṁ bhajāmi.* He is within this universe, but He's also within your heart.

> He's within the atom even. So it is not difficult to find Him out, but you must know the process how to find Him out. And this process is very simple. And the process we are distributing by the order of Śrī Caitanya Mahāprabhu to everyone, without any charge. The process is chant Hare Kṛṣṇa. As soon as you chant Hare Kṛṣṇa, immediately you understand Kṛṣṇa.

In addition to illustrating the purpose of human life – in Ṛṣabhadeva's language, what can and cannot be attained – Śukadeva Gosvāmī describes the Supreme Lord's greatness and His nondifference from everything in the universe. In this regard, while describing Jambūdvīpa in chapter 16, he relays the prayers of outstanding persons such as Lord Śiva, Bhadraśravā, Prahlāda Mahārāja, Lakṣmīdevī, Manu, the goddess of the earth, Hanumān, and Nārada Muni, each of whom worships the Lord in His deity form and glorifies His attractive qualities, unlimited potencies, beauty, and compassionate activities (chapters 17–19). Later, when describing the six inhabited islands of Bhū-maṇḍala (chapter 20), Śukadeva shows that the most exalted demigods (the sun, moon, fire, water, air, and Brahmā, the form of Vedic knowledge), the predominating deities of the universe's life-sustaining elements, derive their potency from the self-effulgent, all-pervading Vāsudeva, the Supreme Personality of Godhead.

Additionally, when Śukadeva describes Sutala, one of the lower planets, he reveals the brilliant gem of surrendered devotion, Bali Mahārāja, as he describes to Mahārāja Parīkṣit the transcendental, independent, self-manifesting nature of *bhakti*.

Near the end of Canto Five Śukadeva describes Anantadeva-Saṅkarṣaṇa and shows the potent, beautiful form and benign nature of the Transcendental Autocrat. He also points to the merciful nature of Kṛṣṇa's holy names, nondifferent from the Lord Himself, even when chanted in a mundane universe with less than pure consciousness.

Conclusion

While describing the universe Śukadeva Gosvāmī repeated the words of devotees, from Dhruva on the topmost polestar to Bali Mahārāja in lowly Sutala. Each expressed a distaste for sense pleasures and an enthusiasm for serving Kṛṣṇa with devotion. Even the wise residents of heaven strongly desire to attain Lord Vāsudeva's mercy – so much so that they long to take birth in Bhārata-varṣa, in the less evolved middle planetary system of Jambūdvīpa, to better facilitate their spiritual advancement.

By seeing the universe with detachment and understanding it in relationship to Kṛṣṇa as His *virāṭ-rūpa,* by hearing of the Lord from His devotees and then following their prayerful example as they worship His *mūrti,* and by noting the Lord's powerful presence in the elements that maintain our existence, sincere souls can see and serve Vāsudeva-Kṛṣṇa even in this dark material world. And in time, our developing pure, devoted consciousness will fully reveal Him as the goal of our lives. This, in brief, is the philosophical thrust Śukadeva Gosvāmī presented as he described the universe in the Fifth Canto of *Śrīmad-Bhāgavatam.*

APPENDIX 2

Excerpt from: *Vedic Cosmography and Astronomy*
Section 3.A (pp. 47–54)

Bhū-Maṇḍala, or Middle Earth

by Richard L. Thompson

The Vedic literature describes the material cosmos as an unlimited ocean situated within a small part of the unlimited spiritual world. Within this ocean there are innumerable universes, or *brahmāṇḍas*, which can be compared to spherical bubbles of foam grouped in clusters. Each of these universal globes consists of a series of spherical coverings and an inner, inhabited portion.

Within the inner region of the *brahmāṇḍa*, the most striking feature is Bhū-maṇḍala, or the earthly planetary system. Bhū-maṇḍala is described in the Fifth Canto of *Śrīmad-Bhāgavatam* as a flat disc with a diameter of 500 million *yojanas*, or 4 billion miles (using 8 miles per *yojana*). The surface of this disc is marked with a series of ring-shaped oceans and islands surrounding a central island called Jambūdvīpa.

The total surface area of our familiar earth planet is some 197 million square miles, and, according to modern information, the total surface area of the sun is about 2.4 million million square miles. In contrast, the total area of Bhū-maṇḍala comes to about 12.6 billion billion square miles. In SB 2.5.40p Śrīla Prabhupāda refers to this as the area of the universe, and it seems that Bhū-maṇḍala is indeed one of the most significant and frequently mentioned features in the Vedic account of the universe. Its size is on the scale of the solar system as a whole, as conceived in modern Western astronomy.

The Fifth Canto gives specific figures for the size, shape, and position of many of the geographic structures of Bhū-maṇḍala. The most striking characteristic of these structures is that although their description employs names for familiar features of earthly geography, such as mountains, oceans, and islands, they are all on the same cosmic scale as Bhū-maṇḍala itself. Thus the smallest mountains on Bhū-maṇḍala mentioned in the *Bhāgavatam* are 2,000 *yojanas*, or 16,000 miles, high. Many mountains are 80,000 miles or even 672,000 miles high. In contrast, the diameter of the earth is about 8,000 miles, and Mount Everest, the highest known mountain, extends about 5.5 miles above sea level. References to such immense sizes are not limited to the Fifth Canto. For example, SB 4.6.32 gives a description of Lord Śiva meditating underneath a banyan tree 800 miles in height and 600 miles in breadth. In SB 8.2.1 we read that Trikuta Mountain, where the elephant Gajendra achieved liberation, is 80,000 miles in length and breadth. This mountain is situated in the ocean of milk, one of the geographical features of Bhū-maṇḍala. In SB 8.7.9 it is pointed out that when Kūrma, the tortoise incarnation of Lord Viṣṇu, was supporting Mandara Mountain during the churning of the milk ocean, His back extended for 800,000 miles (*lakṣa-yojana*), "like a large island." Finally, the Matsya *avatāra*, Lord Viṣṇu's fish incarnation, expanded from an initial small size to a final length of 8 million miles (SB 8.24.44).

Modern scholars tend to reject dimensions such as these as ludicrous exaggerations made by poets who were completely devoid of scientific knowledge. However, even common men in primitive societies can tell that the earthly mountains of our experience have heights of thousands of feet rather than thousands of miles. The highly rational philosophical discussions in the *Bhāgavatam* indicate that it was not written by some kind of mad fanatic who was devoid of common sense. We suggest, therefore, that the descriptions in the *Bhāgavatam* of gigantic sizes refer to an actually existing world that is built on the scale of the solar system and that contains features built on a similar scale.

We will assume that this is the case, and later on we will consider what the relation might be between this world and the earth of our experience. For the present we will give a brief overview of the most significant features of Bhū-maṇḍala. We will do this with the aid of a series of computer-generated illustrations that portray the features of Bhū-maṇḍala as they would appear to an observer approaching Bhū-maṇḍala from a great distance.

In the first view (Fig. 3) we are looking down on the center of Bhū-maṇḍala at an angle of 45 degrees from a distance of some 600 million miles. We can discern five ring-shaped structures surrounding a central region that is too far away to see clearly. Going from the outside in, these are respectively the *dvīpas*, or islands, named Puṣkaradvīpa, Śākadvīpa, Krauñcadvīpa, Kuśadvīpa, and Śālmalīdvīpa. Puṣkaradvīpa has inner and outer radii of 100.4 million and 151.6 million miles, and each successive ring, going inward, is half as wide as the one preceding it. To give an idea of the scale, the distance from the earth to the sun is currently accepted to be 93 million miles.

Figure 26: The plane of Bhū-maṇḍala, looking down at an angle of 45° and at a distance of 600 million miles from the center. (This is the apparent distance if the figure is viewed from about two feet.) The *dvīpas* named Puṣkaradvīpa, Śākadvīpa, Krauñcadvīpa, Kuśadvīpa, and Śālmalīdvīpa, appear as concentric rings, going toward the center from the outside. Plakṣadvīpa and Jambūdvīpa appear to be merged together at the center.

The intervals between the *dvīpas* are occupied by oceans, each of which has the same width as the *dvīpa* that it surrounds. The oceans surrounding the five *dvīpas* we have mentioned are said to be composed respectively of clear water, yogurt, milk, ghee, and liquor. Of course, these substances are celestial counterparts of the corresponding ordinary substances of our day-to-day experience. In Figure 4 we have moved in to a distance of about 150 million miles from the center of Bhū-maṇḍala. Now Krauñcadvīpa, Kuśadvīpa, and

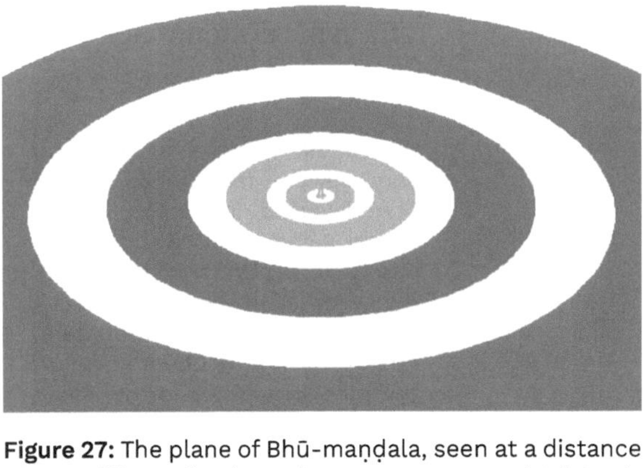

Figure 27: The plane of Bhū-maṇḍala, seen at a distance of 150 million miles from the center. Now Jambūdvīpa is barely visible in the center of the innermost ring, called Plakṣadvīpa.

Śālmalīdvīpa have expanded in apparent size, and the ring of Plakṣadvīpa has become visible within Śālmalīdvīpa. We can also begin to discern the central island of Jambūdvīpa within Plakṣadvīpa. In Figure 5 we have moved in to a distance of 15 million miles, and in Figure 6, at a distance of some 3 million miles, we can obtain a detailed view of Jambūdvīpa.

Figure 28: The plane of Bhū-maṇḍala, seen from a distance of 15 million miles. Jambūdvīpa and the surrounding salt-water ocean are now visible.

Jambūdvīpa is described as a disc-shaped island 100,000 *yojanas*, or 800,000 miles, in diameter. (For comparison, the currently accepted diameter of the sun is 865,110 miles.) The most striking feature of Jambūdvīpa is a central structure called Mount Meru, which is 84,000 *yojanas* high. This structure is generally referred to as a mountain, although it clearly has a unique form quite different from that of a typical mountain. The

Figure 29: A detailed view of Jambūdvīpa from the south. The inverted cone is Mount Sumeru, and on its summit lies Brahmapurī and the residents of eight principal demigods. The smaller mountain ranges form the boundaries of the nine *varṣas*. The arrow indicates the location of Bhārata-varṣa. This picture is rendered according to the descriptions given in the Fifth Canto of the *Śrīmad-Bhāgavatam*.

upper surface of Mount Meru is said to be occupied by Brahmapurī, the city of Lord Brahmā, and by cities belonging to eight other demigods.

Jambūdvīpa is divided into nine regions, or *varṣas*, by a series of mountain ranges. In Figure 7 we see a more detailed view of the central region of Ilāvṛta-varṣa, which contains Mount Meru and is square in shape. To get some idea of the scale of this figure, we should note that the low mountain chain stretching from A to B in the figure is called Gandhamādana and reaches 16,000 miles in height. This is twice the diameter of the earth.

Figure 30: A closer view of Ilāvṛta-varṣa, the central subdivision of Jambūdvīpa. Only those mountains are shown for which dimensions are given in the *Śrīmad-Bhāgavatam*. The mountain ranges stretching from a to b are 16,000 miles in average height, or twice the diameter of the earth.

Of the nine *varṣas* of Jambūdvīpa, eight are described as places of heavenly enjoyment. These are intended for persons who have returned to earth after using up their allotted time on the heavenly planets but who have some remaining pious credits entitling them to enjoy great material opulence. The inhabitants of these *varṣas* are described as living for 10,000 years by earthly calculations, as having the bodily strength of 10,000 elephants, and as having a standard of pleasure like that of the human beings of Tretā-yuga (SB 5.17.12). These regions are also said to contain beautiful gardens that are visited by important leaders among the demigods. The *Bhāgavatam* refers to these eight *varṣas* as *bhauma-svarga*, or the heavenly places on earth (SB 5.17.11), while Śrīla Prabhupāda describes them as "the lower heavenly planets" and contrasts their inhabitants to those of "this earth" (SB 5.17.13p).

The remaining *varṣa* of Jambūdvīpa is called Bhārata-varṣa. It is described as the field of fruitive activities, in which human beings struggle with adverse conditions and elevate or degrade themselves by their actions. Bhārata-varṣa is the southernmost region of Jambūdvīpa, and it is illustrated in Figures 8 and 9, in which we view Jambūdvīpa from the southeast at a lower elevation. In shape, Bhārata-varṣa is a semicircular piece of land bounded on the south by the salt-water ocean and on the north by the Himalayan Mountains. Bhārata-varṣa is the only part of Bhū-maṇḍala at all reminiscent of the earth, and it is frequently identified with either the earth or with India in Śrīla Prabhupāda's books.

Figure 31. Another view of Ilāvṛta-varṣa, with mountains reflecting from the base of Mount Sumeru.

Yet the Himalayas bounding Bhārata-varṣa are described in the Fifth Canto as being 80,000 miles high, and Bhārata-varṣa itself runs some 72,000 miles (9,000 *yojanas*) from north to south. This naturally leads us to ask, What is the relationship between the earth of our experience and Jambūdvīpa and Bhārata-varṣa, as described in the *Śrīmad-Bhāgavatam*?

APPENDIX 3

A Collection of Statements from Śrīla Prabhupāda and His Writings Discussing Vedic Cosmography

a. Lectures, letters, and conversations

Lecture, *Bhagavad-gītā* 2.12, New York, March 9, 1966:

There is a control ... Just like here, in this planet, when you go up you see it as ball, but in this ball there are so many controlling deities here also. President Johnson, prime minister of India, this and that and so many things. But when you go up you see just like a ball.

Lecture (Address), Seattle, September 27, 1968:

The Russian sputnik, circumambulated round this earth in one hour, twenty-five minutes, ... went round the earth for twenty-five time. That means within one hour, twenty-five minutes, the sputnik man saw twenty-five times day and night.

Letter to Dr. Nagendra Babu, Hamburg, August 28, 1969:

Of course, as far as possible I am trying to spread this movement all over the world, and at the present moment my activities are prominent from Hamburg to Tokyo, a distance of 14,000 miles. I think the circumference of the whole earth is 25,000 miles. So this should be covered by some of our Godbrothers so that Lord Caitanya's message may be properly executed.

Lecture (Address), London, September 26, 1969:

And actually, scientifically, it is true that due to the heat of the sun planet all other planets are rotating; otherwise they'll fall down. But they're floating in the air, in the sky, due to this sunlight. Anyone who knows science, he'll admit, "Yes, that's a fact." And sun is the source of all energy in this material world. All this vegetation, all living condition, minerals—there are so many things—this is due to the sun. So sun is the king of all planets, as it is stated in the Vedic literatures.

Letter to Gurudāsa, Los Angeles, May 6, 1970:

Please accept my blessings. I beg to thank you for your letter dated 30th April, 1970, and I have noted the contents carefully. I have also received the list of all the cities and towns in England, and thank you very much. It is my desire that you open a center in each and every one of these places, and that will help to fulfill the prophecy of Lord Caitanya that His Holy Name will be sung in every city and town all over the globe.

Letter to Citsukhānanda, Bombay, March 25, 1971:

I am so glad to hear that you are now in L.A. discussing final plans for opening one center in Mexico City. That is very encouraging news. I am anxious that we establish centers in major cities all over the globe and Mexico City is certainly an important landmark. Both you and your good wife Candrabali can speak Spanish language nicely, so you are certainly qualified for such a mission.

Letter to Kaliya Krishna (John Moffat), Los Angeles, July 8, 1971:

I have given you the spiritual name Kaliya Krishna Das Brahmacari. Kaliya Krishna was a great devotee of Lord Caitanya. Similarly you become a great devotee of Lord Caitanya and help Him to push on this movement to every nook and corner of the globe. Then Caitanya Mahaprabhu will be very much pleased upon you and your life will know success.

Letter to Tribhuvanatha, Brooklyn, July 17, 1971:

Unless we work cooperatively, how this movement will go on? So my request to you all is that you maintain that spirit of cooperation always and help me to deliver Lord Caitanya's message all over the globe.

Lecture (Pandal), Bombay, March 18, 1972:

Therefore this Kṛṣṇa consciousness movement is very, very essential. Very, very essential. I am very glad to inform you that this movement is being especially received in the Western country by the younger generation. I am very much hopeful. I am old man of seventy-six years' age. Now, I can pass away at any moment, but I am confident my disciples, who are mostly Europeans and Americans, they will continue this movement, and I wish there will be considerable change on the face of the globe.

Lecture, *Śrīmad-Bhāgavatam* 1.2.5, Sydney, April 3, 1972:

This whole planet is called Bhārata-varṣa. In the Vedic literature Bhārata-varṣa means this planet. And it is consisting seven islands. That is also mentioned in Vedic literature. Seven islands means Asia, Europe, Africa, North America, South America, Australia, and Oceania.

Jagad-aṇḍa means this universe is a big, I mean to say, volume. Just like *aṇḍa*, egg. Everything, every planet is just like egg. This *brahmāṇḍa*, this universe, is also like egg. So there are many many, many millions of *jagad-aṇḍa*. And in each and every *jagad-aṇḍa, koṭiṣu vasudhādi-vibhūti-bhinnam*, there are innumerable planets also. So these are the information we get from Vedic literature.

Letter to Tamala Krishna, Honolulu, May 16, 1972:

... If the Americans can take up this movement seriously, why not the Bangladesh people? Try to meet with this educated public, and if you can convince them, it will be a great service. We are trying to unify the whole population of this globe by this cultural movement. There are blessings of Lord Caitanya in this connection, and if we try sincerely it will be successful.

Morning walk, Los Angeles, May 13, 1973:

Paramahaṁsa: For example in the Bible they quote that the world is flat. So therefore when they found out the world is round, they say, "Oh, scripture is false."

Prabhupāda: Yes.

Paramahaṁsa: Scientists, they have proven the truth that the world is round. Therefore scientists ...

Prabhupāda: Yes. That must be, therefore the scriptures must be transcendental. Nonsense scripture, and people become nonsense. So except Vedic literature, all nonsense scripture. They are not scripture. Manufactured. This Bible was manufactured by the saint, this saint, that, according to their imagination. It was not spoken by Lord Jesus Christ. What was spoken by Jesus Christ, that they ignore: "Thou shalt not kill." They kill. Nobody is following Christian principle; neither Bible is perfect. But that, if we say, we will be shot. (laughter) ...

Prabhupāda: The earth was flat. They believed that the world is flat. So how much imperfect knowledge they have got. So imperfect knowledge, how long it can go? Just like we are going to challenge all these rascals that life is grown out of matter. We are going to challenge. It is not a fact. So how long you can cheat people? For hundred, two hundred, thousand years, but you cannot cheat for all the time.

Lecture, *Bhagavad-gītā* 1.24-25, London, July 20, 1973:

If you simply drive your plane on the western side, you'll never get night. You just round about. Yes. Just like it is materially possible. You drive your plane simply on the western side. Start your plane in the morning and go to the western side and don't stop it. You go on for three, four, as many days as you like. You'll never get night.

Lecture, *Bhagavad-gītā* 1.43, London, July 30, 1973:

Prabhupāda: ... So Brahmā means the manager of one *brahmāṇḍa*, universe. And there are millions of Brahmās, millions, trillions. They are numberless. *Ananta-koṭi. Yasya prabhā prabhavato jagad-aṇḍa-koṭiḥ* [Bs. 5.40]. *Koṭi. Koṭi* means unlimited. *Jagad-aṇḍa. Jagad-aṇḍa* means universe. *Brahmāṇḍa* or *jagad-aṇḍa. Aṇḍa. Aṇḍa* means it is egg-shaped, round, egg-shaped. Therefore it is called *aṇḍa, brahmāṇḍa. Bhū-gola. Gola* means round. I have heard that before the science, the people were under the impression that this world is square. Is it not?

Devotees: Flat.

Prabhupāda: Flat. Flat, yes. But in the Vedic *śāstra*, millions of years ago it is mentioned: *bhū-gola. Gola* means round. Just see. And these rascals say that formerly people were not so intelligent. They are intelligent because they are thinking that this world is flat. And those who have spoken millions of years ago, "It is round," they will have less intelligence. Just see. *Bhū, gola. Gola* means round. *Bhū-gola.* Similarly, *jagad-aṇḍa. Aṇḍa* means round, just like egg. *Aṇḍa* means egg. *Jagad-aṇḍa.* This universe is egg-shaped. And we can see also, the sky is round. This is the wall of this universe.

Lecture, *Bhagavad-gītā* 13.8–12, Bombay, September 30, 1973:
All the planets and the universes, they are egg shaped. Therefore they are called *aṇḍa, jagad-aṇḍa, brahmāṇḍa.*

Morning walk, Los Angeles, December 9, 1973:
Yes, world is round. That is fact. Goloka. In Vedic literature it is *bhū-gola, jagad-aṇḍa.* These words are there. We can see also it is round, *jagad-aṇḍa.* The universe is round. And Goloka. Or *Bhū-gola. Bhū-gola*, the earth is round. So in the Vedic literatures ... Therefore their knowledge is also imperfect because they do not refer to the Vedic literatures. It is already there. *Bhū-gola. Bhū* means the earth; *gola* means round. It is already there. And the geography's called, according to Sanskrit, it is called *Bhū-gola.* Long, long ago, before Galileo.

Lecture, *Śrīmad-Bhāgavatam* 1.16.17, Los Angeles, January 12, 1974:
And again, if you start your plane in sunlight, in the morning, in daylight, so in the morning, and go to the western side, you will find never night. You will find never night, always light, always light. Sometimes we have got experience. We start from a place, say, at ten o'clock, and going western side. Then, after few hours, we see it is still ten o'clock, and the light is there. This is our practical experience.

Lecture, *Bhagavad-gītā* 7.5, Bombay, February 20, 1974:
The spiritual force behind is moving the material world. *Yayedaṁ dhāryate jagat. Jagat* means *gacchati iti jagat*, which is moving or going forward. Everything is moving. All these planets are moving. Your earthly planet is also moving—within twenty-four hours covering 25,000 miles, day and night. The whole material planets, earthly planets, you have 25,000 miles, and this is rotating. Similarly every planet is rotating. The sun is also rotating.

Lecture, *Śrīmad-Bhāgavatam* 3.26.3, Bombay, December 15, 1974:
Just like on account of the sunshine all the planets are resting and rotating. That is scientific. Due to the heat of the sunshine, all the planets in the sky, they are rotating. It is due to sunshine. Similarly,

this *brahmāṇḍa*, this universe, not only one universe, but millions of universes, they are also rotating in the *brahmajyoti*.

Lecture, *Śrīmad-Bhāgavatam* 3.26.16, Bombay, December 25, 1974:

The United Nations is working so hard, but they cannot make that the whole world belongs to all living being. Then immediately it will be ... All questions will be solved. I was discussing this evening: there is so much land still uncultivated. And if the overpopulated people are allowed to go there and cultivate and grow their food grains, ten times of the living entities on this face of the globe can be fed without any difficulty. *Annād bhavanti bhūtāni* (BG 3.14).

Lecture, *Bhagavad-gītā* 16.8, Tokyo, January 28, 1975:

Now, as we see in the street that the cars are moving in high speed but they are within the orbit, within the line, demarcation of line, white line or yellow line, so there is some brain, there is some management, everything is there, Similarly, all these planets, they are rotating with high speed.

Just like this planet. It is rotating 25,000 miles in twelve hours. Is it not? The circumference of this earth is 25,000 miles and ... Yes, day and night, twenty-four hours. Almost one thousand miles per hour ...

[T]he earth is running at one thousand miles per hour, but we cannot understand. The arrangement is so nice. The perfection that it is ... we cannot understand it. It is practical. We see the morning, day, coming. That means earth is moving. When the aeroplane moves also, there are so many jerking, those sounds. They're all imperfect. But here you see that such perfect arrangement, it is moving one thousand miles per hour, and there is no jerking. There is nothing of the sort. We are thinking, "We are sitting in the same place." And there is no brain?

Morning walk, Mauritius, October 2, 1975:

Brahmānanda: In the Bible there are so many things that were ... about the world that were mistakes, and then, by their scientific knowledge, they found the real fact. Like they say the earth was flat...

Prabhupāda: Bible may be wrong, but we are talking generally. So if you are scientist, you can say the right thing. But why should you deny God?

Brahmānanda: Well, if the book of God has made mistake, then we don't have any faith...

Prabhupāda: Book of God ... Then throw it. Kick it out. But I say that you are scientist. How you can deny God? That is our propaganda. The so-called book of God, you throw away. But you are scientist. How can you deny the existence of God? Why you are misleading? Bible misleading, that's all right. Why you are misleading? That is our point.

Lecture, *Śrīmad-Bhāgavatam* 3.28.21, Nairobi, November 1, 1975:

... *Jagad-aṇḍa* means ... *Aṇḍa*, it is round. *Aṇḍa* means egg shape. We don't say it is flat. *Śāstra* never says, "Flat." Then you correct, "No, no, it is not flat. It is round." This is *śāstra*. Long, long years, in the beginning, it was called round, *aṇḍa*. Goloka, *go*, Goloka, Bhūrloka. *Goloka* means round. Goloka Vṛndāvana. Everything is round ...

Room conversation, Washington, DC, July 5, 1976:

Prabhupāda: So long. As soon as there is five thousand years, they will correct. The Christians, they used to say that the world is flat. So what is the value of their words? There is no value. You do not try to compromise with others.

Letter to Madhava, Vṛndāvana, October 3, 1976:

... Some portion of the earth is flat. When you stand in any place you see flat, so for us to some extent it appears flat, but it is round ... So whether Sukadeva Goswami or the modern scientists are right or wrong, it is nothing interesting to us. We want to go back to home, back to Godhead – *yanti mad-yajino 'pi mam.*

Room conversation, Vṛndāvana, June 28, 1977:

Puṣkara: Śrīla Prabhupāda, previously we painted in the art department ... It's described that Varāha lifted the earth, and the earth was a globe, and we showed also a globe of the earth. How does that relate to this? Previously, when we painted, we showed the earth a ball. So now the artists will be very confused. How it fell in the Garbha Ocean as a ball?

Yaśodānandana: It depends on what we mean by earth. The Western conception of earth is just five continents and a few oceans, but according to *Bhāgavatam*, earth means Jambūdvīpa, because earth is connected with Jambūdvīpa.

Puṣkara: So whole Jambūdvīpa fell.

Bhakti-prema: *Bhāgavata* describes the height of Himalayas, eighty thousand miles.

Prabhupāda: No, about this earth globe.

Room conversation, Vṛndāvana, July 5, 1977:

Bhakti-prema: It is written the world ... The earth is round and flat.

Prabhupāda: Hmm?

Bhakti-prema: Earth is round and flat both, together.

Prabhupāda: Yes.

Bhakti-prema: We can name it Esoteric Geography

Prabhupāda: Journal?

Bhakti-prema: Esoteric.

Prabhupāda: Yes, esoteric I know. Journal?

Bhakti-prema: Geography [indistinct] ... qualification.

Prabhupāda: Yes. Esoteric and exoteric.

Tamāla Krishna: We're drawing according to the *Bhāgavatam*.

Prabhupāda: We're not conceiving it. It is already there.

Bhakti-prema: Then it is conceived, he says.

Prabhupāda: Conceived not by me.

Tamāla Krishna: But not by us. By the Unlimited.

Prabhupāda: I am imperfect. That is the difference.

Bhakti-prema: Yes. It is conceivable by you.

Tamāla Krishna: By Kṛṣṇa.

Prabhupāda: By Śukadeva ... Not even Śukadeva Gosvāmī. He says *śuśruma*.

Tamāla Krishna: "I heard."

[For full transcript see: https://vedabase.io/en/library/transcripts/770705dcvrn/]

Discussion with Śyāmasundara:

And there is Varāha incarnation, He picked up the whole earth on the tusk. How much big the Varāha animal was to show that it can pick up the whole earth, earthly planet just like a ball. Because your senses are imperfect. You cannot see. You cannot search out. Have you searched out all the earthly layers or the 25,000 miles everywhere? That is not possible for you. The whole earthly planet ... circumference is 25,000 miles ...

[Philosophy Discussion on Charles Darwin,
Los Angeles, May 1972]

b. *Śrīmad-Bhāgavatam*

SB 1.6.31, Purport:

Śrī Nārada Muni could enter all these planets in both the material and spiritual spheres without restriction, as much as the almighty Lord is free to move personally in any part of His creation.

SB 1.7.10, Purport:

As far as the mystic powers are concerned, they are eight in all (such as to become the minutest in form, to become weightless, to have anything one desires, to lord it over the material nature, to control other living beings, to throw earthly globes in outer space, etc.). These mystic powers are mentioned in the *Bhāgavatam*.

SB 1.7.30, Purport:

The radiation of atomic energy is very insignificant in comparison to the heat produced by a *brahmāstra*. The atomic bomb explosion can at utmost blow up one globe, but the heat produced by the *brahmāstra* can destroy the whole cosmic situation. The comparison is therefore made to the heat at the time of annihilation.

SB 1.14.38, Purport:

In the heavenly planets, the state assembly house called Sudharmā was deserving of the dignity of the best of the demigods. Such an assembly house is never meant for any state on the globe because the human being on the earth is unable to construct it, however far a particular state may be materially advanced. But during the time of Lord Kṛṣṇa's presence on the earth, the members of the Yadu family forcibly brought the celestial assembly house to earth and placed it at Dvārakā.

SB 1.18.35, Purport:

There are many false incarnations after the departure of Lord Kṛṣṇa from the face of the globe, and they are misleading the innocent public by accepting the spiritual obedience of the general mass of people to maintain false prestige.

SB 2.4.8, Purport:

Even those who are constantly engaged in sinful acts are all corrigible to the standard of perfect human beings if they take shelter of the devotees of the Lord. Jesus Christ and Muhammad, two powerful devotees of the Lord, have done tremendous service on behalf of the Lord on the surface of the globe.

SB 2.7.1, Purport:

Only under certain conditions do the planets float as weightless balls in the air, and as soon as these conditions are disturbed, the planets may fall down in the Garbhodaka Ocean, which covers half the universe. The other half is the spherical dome within which the innumerable planetary systems exist. The floating of the planets in the weightless air is due to the inner constitution of the globes, and the modernized drilling of the earth to exploit oil from within is a sort of disturbance by the modern demons and can result in a greatly harmful reaction to the floating condition of the earth.

SB 2.8.15, Translation:

O best of the *brāhmaṇas*, please also describe how the creation of the globes throughout the universe, the four directions of the heavens, the sky, the planets, the stars, the mountains, the rivers, the seas and the islands, as well as their different kinds of inhabitants, takes place.

SB 2.9.5, Purport:

In the beginning of *Śrīmad-Bhāgavatam* it has already been said that the Supreme Absolute Truth exists in His own abode without any touch of the deluding energy. Therefore the kingdom of God is not a myth but factually a different and transcendental sphere of planets known as the Vaikuṇṭhas. This will also be explained in this chapter.

SB 3.17.5, Purport:

When there are natural disturbances like blowing cyclones, too much heat or snowfall, and uprooting of trees by hurricanes, it is to be understood that the demoniac population is increasing and so the natural disturbance is also taking place. There are many countries on the globe, even at the present moment, where all these disturbances are current.

SB 3.17.28, Translation:

You are the guardian of an entire sphere and a ruler of wide fame. Having crushed the might of arrogant and conceited warriors and having conquered all the Daityas and Dānavas in the world, you once performed a Rājasūya sacrifice to the Lord.

SB 3.21.2 Purport

Śrīmad-Bhāgavatam is also a history of the great rulers of different parts of the universe. In this verse, the names of Priyavrata and Uttānapāda, the sons of Svāyambhuva Manu are mentioned. They ruled this Earth, which is divided into seven islands. These seven islands are still current, as Asia, Europe, Africa, America, Australia and the North and South Poles.

SB 3.21.52–54, Translation:

If you did not mount your victorious jeweled chariot, whose mere presence threatens culprits, if you did not produce fierce sounds by the twanging of your bow, and if you did not roam about the world like the brilliant sun, leading a huge army whose trampling feet cause the globe of the earth to tremble, then all the moral laws governing the *varṇas* and *āśramas* created by the Lord Himself would be broken by the rogues and rascals.

SB 3.22.31, Translation:

Manu spread a seat of *kuśas* and *kāśas* and worshiped the Lord, the Personality of Godhead, by whose grace he had obtained the rule of the terrestrial globe.

SB 3.23.41, Purport:

The planets occupied by the demigods are restricted to their own orbits, but Kardama Muni, by his yogic power, could travel all over the different directions of the universe without restriction. The living entities who are within the universe are called conditioned souls; that is, they are not free to move everywhere. We are inhabitants of this earthly globe; we cannot move freely to other planets.

SB 3.23.43, Translation:

After showing his wife the globe of the universe and its different arrangements, full of many wonders, the great *yogī* Kardama Muni returned to his own hermitage.

SB 3.23.43, Purport:

All the planets are here described as *gola*, round. Every planet is round, and each planet is a different shelter, just like islands in the great ocean. Planets are sometimes called *dvīpa* or *varṣa*. This earth planet is called Bhārata-varṣa because it was ruled by King Bhārata.

SB 4.4.24, Purport:

The real *yoga* process accepts the transmigration of the soul from one planet to another or one body to another; and it appears from this incident that Satī wanted to transfer her soul to another body or sphere.

SB 4.16.20, Translation:

This King, being uniquely powerful and heroic, will have no competitor. He will travel around the globe on his victorious chariot, holding his invincible bow in his hand and appearing exactly like the sun, which rotates in its own orbit from the south.

SB 4.21.48, Translation:

All the saintly *brāhmaṇas* thus addressed Pṛthu Mahārāja: O best of the warriors, O father of this globe, may you be blessed with a long life, for you have great devotion to the infallible Supreme Personality of Godhead, who is the master of all the universe.

SB 5.7.1, Translation:

Śukadeva Gosvāmī continued speaking to Mahārāja Parīkṣit: My dear King, Bhārata Mahārāja was a topmost devotee. Following the orders of his father, who had already decided to install him on the throne, he began to rule the earth accordingly. When Bhārata Mahārāja ruled the entire globe, he followed the orders of his father and married Pañcajanī, the daughter of Viśvarūpa.

SB 5.13.23, Purport:

The four Kumāras walked everywhere in the guise of five-year-old boys, and similarly there are many *brāhmaṇas*, knowers of Brahman, who traverse the globe either as young men, children or *avadhūtas*.

SB 5.16.4, Translation:

... O King, I shall nevertheless try to explain to you the principal regions, such as Bhūloka [Bhū-golaka], with their names, forms, measurements and various symptoms.

SB 5.20.38, Purport:

sa tu lokālokas tu bhū-golakasya bhū-sambandhāṇḍa-golakasyety arthaḥ; sūryasy eva bhuvo 'py aṇḍa-golakayor madhya-vartitvāt kha-golam iva bhū-golam api pañcāśat-koṭi-yojana-pramāṇaṁ tasya turīya-bhāgaḥ ...

The Sanskrit mentioned above is taken from Viśvanātha Cakravartī's commentary. Prabhupāda describes this commentary as follows: "However, the technical terms used in the astronomical calculations given by the *Jyotir Veda* are difficult to translate into English. Therefore to satisfy the reader, we may include the exact Sanskrit statement given by Śrīla Viśvanātha Cakravartī Ṭhākura, which records exact calculations regarding universal affairs."

SB 5.21.2, Translation:

As a grain of wheat is divided into two parts and one can estimate the size of the upper part by knowing that of the lower, so, expert geographers instruct, one can understand the measurements of the upper part of the universe by knowing those of the lower part. The sky between the earthly sphere and heavenly sphere is called *antarikṣa*, or outer space. It adjoins the top of the sphere of earth and the bottom of that of heaven.

SB 8.19.5, Translation:

It was in your dynasty that Hiraṇyākṣa was born. Carrying only his own club, he wandered the globe alone, without assistance, to conquer all directions, and no hero he met could rival him.

SB 10.8.37–39, Translation:

When Kṛṣṇa opened His mouth wide by the order of mother Yaśodā, she saw within His mouth all moving and nonmoving entities, outer space, and all directions, along with mountains, islands, oceans, the surface of the earth [*bhū-golam*] ...

c. *Śrī Caitanya-caritāmṛta*

CC *Adi* 5.20, Purport:

Therefore, since the living being is spiritual in constitution, he can be happy only in the spiritual sky, where there are unlimited spiritual spheres called Vaikuṇṭhas.

CC *Adi* 5.22, Purport:

Kṛṣṇaloka is the topmost planet in the spiritual sky, and below it are innumerable spheres, a description of which can be found in *Śrīmad-Bhāgavatam*. In the beginning of Lord Brahmā's self-realization he was shown a transcendental vision of the Vaikuṇṭha spheres by the grace of Nārāyaṇa.

CC *Adi* 5.119, Purport:

The Lord of Śvetadvīpa expands Himself as Śeṣa Nāga, who sustains all the planets upon His innumerable hoods. These huge global spheres are compared to grains of mustard resting on the spiritual hoods of Śeṣa Nāga.

The scientists' law of gravity is a partial explanation of Lord Saṅkarṣaṇa's energy. The name "Saṅkarṣaṇa" has an etymological relationship to the idea of gravity. There is a reference to Śeṣa Nāga in *Śrīmad-Bhāgavatam* (5.17.21), where it is said:

yam āhur asya sthiti janma-saṁyamaṁ
tribhir vihīnaṁ yam anantam ṛṣayaḥ
na veda siddhārtham iva kvacit sthitaṁ
bhū-maṇḍalaṁ mūrdha-sahasra-dhāmasu

"O my Lord, the hymns of the *Vedas* proclaim that You are the effective cause for the creation, maintenance and destruction. But in fact You are transcendental to all limitations and are therefore known as unlimited. On Your thousands of hoods rest the innumerable global spheres, like grains of mustard so insignificant that You have no perception of their weight." The *Bhāgavatam* further says (5.25.2):

yasyedaṁ kṣiti-maṇḍalaṁ bhagavato
'nanta-mūrteḥ sahasra-śirasa ekasminn
eva śīrṣaṇi dhriyamāṇaṁ siddhārtha iva lakṣyate

"Lord Anantadeva has thousands of hoods. Each sustains a global sphere that appears like a grain of mustard."

CC *Adi* 5.120, Purport:

... The *Laghu-bhāgavatāmṛta* gives the following description: "The Saṅkarṣaṇa of the second group of quadruple forms appears as Rāma, taking with Him Śeṣa, who bears the global spheres. There are two features of Śeṣa. One is the bearer of the globes, and the other is the bedstead servitor. The Śeṣa who bears the globes is a potent incarnation of Saṅkarṣaṇa, and therefore He is sometimes also called Saṅkarṣaṇa. The bedstead feature of Śeṣa always presents himself as an eternal servitor of the Lord."

CC *Adi* 17.19, Purport:

One day Caitanya Mahāprabhu began to cry out, "Śūkara! Śūkara!" Thus crying out for the boar incarnation of the Lord, He assumed His form as the boar incarnation and got up on the shoulders of Murāri Gupta. He carried a small *gāḍu,* a small waterpot with a nozzle, and thus He symbolically picked up the earth from the depths of the ocean, for this is the pastime of Lord Varāha.

CC *Adi* 7.164, Purport:

... Now that the International Society for Krishna Consciousness has taken up this task of preaching the cult of Lord Caitanya, its members should not only construct temples in every town and village of the globe but also distribute the books that have already been written and further increase the number of books. Both distribution of books and construction of temples must continue side by side in parallel lines.

d. Other works by A. C. Bhaktivedanta Swami Prabhupāda

Kṛṣṇa: The Supreme Personality of Godhead, Chapter 85:

"I know that neither of You is my son; both of You are the original chief and progenitor, the Personality of Godhead, the Puruṣa with *pradhāna*. But You have appeared on the surface of this globe to minimize the burden of the world by killing the *kṣatriya* kings who are unnecessarily increasing their military strength. You have already informed me about this in the past. My dear Lord, You are the shelter of the surrendered soul, the supreme well-wisher of the meek and humble. I am therefore taking shelter of Your lotus feet, which alone can give one liberation from the entanglement of material existence."

Light of the Bhāgavata, Verse 3, Purport:

Water is also required very liberally for many other purposes. Thus the Lord has preserved water on three fourths of the globe and has made it salty to preserve it. Salty water does not decompose, and that is the arrangement of providence.

Light of the Bhāgavata, Verse 21, Translation:

The wind carries the clouds to different parts of the globe, and the clouds distribute rains, to the satisfaction of the people in general, just as rich kings and merchants distribute their accumulated wealth, inspired by religious priests.

Mukunda-mālā-stotra (*mantras* 1 to 6):

By His spiritual energy Śeṣa Nāga sustains all the planetary globes upon His invisible hoods. Śeṣa Nāga is popularly known as Saṅkarṣaṇa, or "that which keeps balance by the law of magnetism." In the scientific world this feature of the Lord is referred to as the law of gravitation, but factually this law, which keeps all the planets floating in space, is one of the energies of the Lord.

Message of Godhead, Chapter 2:

If we examine human affairs in the light of the caste system as created by the Personality of Godhead, surely we can visualize the four social orders functioning in every part of the world. In every part of the globe, wherever there is human habitation, there are some persons who have the qualifications of *brāhmaṇas*, and there are others who have the qualifications of *kṣatriyas*, *vaiśyas*, and *śūdras*.

APPENDIX 4

An Introduction to Śrī Va□śīdhara and His Sanskrit Commentary

by S. E. Kreitzer

Śrī Vaṁśīdhara Śarman is a highly respected commentator within a legacy of Vaishnava *ācāryas*, though perhaps not as well-known to English speaking audiences until Richard L. Thompson published an account of Vaṁśīdhara's analysis of *Srimad-Bhāgavatam* 5.20.38, in an appendix to his own work, *Vedic Cosmography and Astronomy* (1989). Śri Vaṁśīdhara's appraisal of this Fifth Canto verse suggests a complexity of perspectives brought to bear in an examination of the apparent dichotomy between the mathematically quantifiable visible universe, and the descriptions of the cosmos located within Purānic literature. Both were well known to learned *paṇḍitas* within the Vedic cultural tradition.

A draft of the full commentary, located in Thompson's personal files, had been drawn primarily from the work of two well-regarded Sanskrit scholars, V. N. Jha of the University of Pune, and V. V. Gangal from the Ananthacharya Indological Research Institute. Thompson contacted them while securing research material for his book, and he utilized a number of lengthy excerpts from this draft based on their work, for use in the Appendix chapter "Vaṁśīdhara on Bhū-maṇḍala and the Earth Globe." The balance of the full commentary is taken from the draft manuscripts presently located with the Richard L. Thompson Archives. (Box 5, Folder 02, Richard L. Thompson Archives, 12834 NW 151 Rd., Alachua, FL 32615, USA. Email: rlthompsonarchives@ivs.edu.)

Along with Vaṁśīdhara's commentary to this verse, this Appendix includes autobiographical material written by Śri Vaṁśīdhara which he added at the conclusion of his work, *Bhāvārtha-dīpikā-prakāśa*. The English translation is by Krishna Swarūpa dāsa, a Vaishnava *paṇḍita* originally from Nepal who had previously worked at the Bhaktivedanta Institute on research projects for Richard L. Thompson. Most significantly, Śri Vaṁśīdhara's autobiographical content illustrate the sentiments as a maturely realized *sādhu*, whose well-established writings are frequently mentioned alongside the works of respected Vaishnava *ācāryas* of the highest caliber.

Following that is an essay offering a historiographical account of Śri Vaṁśīdhara's biographical information as appreciated by scholars prior to Chaukhamba Sanskrit Pratishthan publishing his full Sanskrit commentary in 1993. Whereas a previous consensus generally gravitated toward placing Vamsidhara around the 17th century, information included with the publication of his Sanskrit commentaries in the 1990s appears to solidly place him in the second half of the 19th century. While on one hand, Śri Vaṁśīdhara may well be considered among the last of the great Vaishnava Sanskrit language commentators, perhaps he could also be considered among the first to write with awareness of the contemporary flow of discourse as well.

a. Translation of Śrī Va□śīdhara's purport to *Śrīmad-Bhāgavatam* 5.20.38

What follows is a copy of a full translation of Śrī Vaṁśīdhara's commentary for *Śrīmad-Bhāgavatam* 5.20.38 drawn from research files held by the Richard L. Thompson Archives. This translation drew upon the work of two accomplished Sanskrit scholars, V. V. Gangal and V. N. Jha, with excerpts from their translations published in an Appendix chapter on Vaṁśīdhara's commentary included in Thompson's pioneering work, *Vedic Cosmography and Astronomy* (1989). The balance of the translation presented here was drawn from copies of the original manuscripts held by the Richard L. Thompson Archives in Box 5, Folder 02. For more information on these texts, please contact: Richard L. Thompson Archives, 12834 NW 151 Rd., Alachua, FL 32615, USA. Email: info@richardlthompson.com.

The translation to the verse from *Śrīmad-Bhāgavatam* 5.20.38 is taken from Bhaktivedanta Book Trust edition of the *Śrīmad-Bhāgavatam* Fifth Canto by A. C. Bhaktivedanta Swami Prabhupāda.

Srimad-Bhāgavatam *5.20.38*
Translation by A. C. Bhaktivedanta Swami Prabhupāda

> Learned scholars who are free from mistakes, illusions and propensities to cheat have thus described the planetary systems and their particular symptoms, measurements and locations. With great deliberation, they have established the truth that the distance between Sumeru and the mountain known as Lokāloka is one fourth of the diameter of the universe—or, in other words, 125,000,000 *yojanas* [1 billion miles].

English Translation of Śrī Vaṁśīdhara's Commentary on Srimad-Bhāgavatam *5.20.38*

The expression "*sa tu*" stands for Lokāloka mountain. "*Bhūgolasya*" means "of the egg-like sphere [the inner earth sphere of the universal egg] connected with the earth." Like the Sun, the Earth is also midway between two egg-like [hemi]spheres. Like the space-sphere, the earth-sphere measures 500 million *yojanas*. "*turīya-bhāgaḥ*" means that one-fourth of this, or 125 million *yojanas*, is the radius [?] of Lokāloka mountain.

The Earth should be known, however, as extending 496,600,000 *yojanas* [or 50 *crores* minus 34 *lakhs*]. Thus, from the middle of Mt. Sumeru to the middle of Mānasottara Mt. is 15,750,000 *yojanas*; from there to the far shore of the sweet-water ocean is 9,600,000 *yojanas*. After another 15,750,000 *yojanas* is the golden land, which is uninhabited [see SB 5.20.35]. Thus, on one side the distance between Mt. Sumeru and the end of the inhabited land, or Loka, is 41,100,000 *yojanas* [i.e. 411 = 2×157.5 + 96]. On the opposite side it is the same. From the inhabited land [Loka] to Lokāloka Mountain, the distance is 82.2 million *yojanas*. This is the extent of this outside portion on one side of Mt. Sumeru, and it is also the same on the other side. This sums up the total extent of the region within Lokāloka Mountain. The extent of the Aloka-varṣa outside of the Lokāloka mountain is: On one side the outside region of Lokāloka measures 125 million *yojanas*, and on the other side of the earth it is the same. Thus the earth, with its islands and oceans, measures 496,600,000 *yojanas* [i.e. 2 × (41.1 + 82.2 + 125) million *yojanas*].

Thus, a gap of 1,700,000 *yojanas* exists in all directions between the earth and the inner sphere of the universal egg. Due to this gap it is meaningful for the earth to be supported by Lord Śeṣanāga and to be held steady by the elephants of the four directions. But if the earth were 500 million *yojanas* across, then it would contact the universal egg-sphere on all sides, and this would be meaningless. The sudden immersion of the earth in the Cākṣuṣa Manu epoch and its lifting by Lord Varāha would also be difficult to explain. This and other difficulties should be considered.

This is also considered by Śrī Nīlakaṇṭha in the *Bhīṣma-parva* of the *Mahābhārata*: "Here, on account of non-contradiction between scriptures and experience, this conception of the earth appears correct."

For example, the periphery of a square measuring one *hasta* [or cubit] on a side is twenty-four *aṅgulis* [or fingers] and the circumference is ninety six *aṅgulis*. The diagonal is somewhat shorter. The extent of Jambhūdvīpa is declared to be 18,600 *yojanas*; here we conceive of this as the perimeter, and thus each side is 4,650 *yojanas*. This is the diameter of the square, and according to *Śulbasūtras* the diagonal measures 6576 *yojanas* [by the "Pythagorean theorem"]. The earth-square is encircled by four oceans and its corners are directed towards the cardinal points; the meeting-place of the two oceans in the southern quarter is at the bridge of Rāma. Likewise, pairs of oceans meet at the corners in the other directions. The circumference of the earth according to astronomers, namely 4967 *yojanas*, is the same as the diameter of our rectangle, or 4650 *yojanas*.

The difference of these two numbers, namely 317 *yojanas*, can be reconciled by considering another system of measures described by Kātyāyana and others. According to this, one *hasta* equals one-fifth of the height of the sacrificer who stands stretching his hands up. Due to this difference in the measurement of a *hasta* there may be a difference (more or less) in the length of a *yojana*. This is how one should solve the problem of the difference.

Likewise, although the earth is rectangular, the course of the sun circumambulating (clockwise) the Meru Mountain stationed in its middle would be circular. Thus, there is no necessity of conceiving a pair of suns as is done by the Buddhists. All else that is not contrary to logic as considered by astronomers may be followed.

When it is stated in the *Puraṇās*, that the measure of the earth is 500 million *yojanas*, this can be answered by citing the maxim, "The things which are beyond imagination, one should not try to establish by logic or argument." Since there are contradictions between these doctrines and current opinions, it is better to resort to the theory of inexplicability [*anirvacanīya*]. Or, whatever measure is mentioned in the Puraṇās, one should take l/20th of that and thus, in place of 500 million, one should accept 25 million *yojanas* as the measure of the earth. In place of one *lakh*, 5,000 *yojanas* should be accepted as the diameter of Jambūdvīpa; in place of 9,000, 450 *yojanas* should be accepted as the measure of India (Bhārata).

According to this view, as given by the statements of the Vaishnavas that we shall present, Jambūdvīpa is like a lotus of four petals. Its outer circumference is 18,600 *yojanas*; its central diameter is 3300 *yojanas*. So virtually [lit. as a result] the diameter may be understood to be 5,000 *yojanas*. According to this view, since the Himalaya Mountains touch the eastern and western oceans, this Bhārata-varṣa becomes a triangle. Therefore, our experience and the popular saying, "the earth is a triangle," go together.

Otherwise, if Bhārata-varṣa were of the form of a bow, then the mid-earth longitude starting from Laṅkā, proceeding by the path of the bridge of Rāma (Rāmasetu), will go away from the line mentioned in the following verse: "This is the central line of the earth: the city of Rāksasas, Kāñcī, the daughter of God, then the white mountain, Paryalivatsagulma, the city by name of Ujjayinī, Gargarata, and Kurukṣetra." With this line Rāmasetu will come closer to Kurukṣetra and Dvārakā will go far from it. But it is our direct experience that Dvārakā is near and Rāmasetu is at a far distance. The conclusion is that it is better to regard Bhārata-varṣa [India] as triangular. In this way, Nīlakaṇṭha discussed this in the *Mahābhārata* and removed the contradictions.

What's more, according to the *Smṛtisāra* and other sources, "Four times a *hasta* of 24 fingers breadth is called a *daṇḍa* [= 2 yards]. Two thousand *daṇḍas* equal 1 *krośa* [about 2 miles] and four such *krośas* equals 1 *yojana* [about 8 miles]." Also "a *yojana* is (measured by) thirty-two thousand cubits (*hastas*)." This is corroborated by the statement of the *Saṁhitā*, "The *yojana* of the earth is thirty-two thousand *hastas*," and also "A *hasta* is said to be twenty-four fingers." And as there are differences in the breadth of the fingers of a newly-born baby, a boy, and a young man, there must thereby be differences in the size of a *krośa* and a *yojana*. Or, as the dictionary states that "palm" and "hand" are synonymous, so *hasta*, *pañcaśākha* [five-fingered limb], and *kara* are synonymous measures. This means that one *hasta* stands for the measure of the hand above the wrist, i.e., about 8 or 9 *aṅgulas*. The intention to give is accepted with a bent (i.e. cupped) hand in such statements as "gifts may be given in the hand of a *brāhmana*." There too the term "*hasta*" is used, and with the contraction of fingers it also becomes four-finger-long. Of a child it becomes still less. And so by counting *yojanas* by "*hastas*" thus and measuring differently, the contradictions among books of astronomy, and the *Purāṇas* and *Itihāsas*, could somehow be avoided.

Not indeed has the *jyotiḥśāstra*, or science of luminaries, started contravening the Purāṇic statement that "Vyāsa is Nārāyaṇa himself." Nor could Vyāsa also have proceeded in contravention of the science of luminaries, which is the very eye of *Veda*, as expressed in the statement, "Astronomy is declared to be the eye [of the *Veda*]." Therefore, at different places, statements as to *yojanas* may be inferred to mean those various measures of finger, hand, etc. Moreover, it appears that Vyāsadeva himself speaks contrary to astronomy, as it were, in order to curb the tendency on the part of *asuras* toward the study of the *śāstras*. But truly speaking, he is not doing so. Otherwise, it may be contemplated by the well-intentioned that there would be darkness [i.e. ignorance] as to *Veda* on the part of Vyāsa. This is the proper understanding.

Well, then how can one explain the contradiction between the *Bhāgavata* and the *Jyotiḥśāstra*, or astronomical science? In the *Bhāgavata*, Jambūdvīpa is said to measure 100,000 *yojanas*, whereas the astronomical science states the entire earth to be measuring only 5,000 *yojanas*. The solution is given in the *Goladarśa*. According to that text some brief explanation is given below:

The earth has two forms. One is the particular [*viśeṣa*] form of big measure, and the other is the smaller, spherical form given in the *Jyotiḥśāstra*. In this regard Parīkṣit asked Śrī Śuka, the great *yogin*, and he replied [in SB 5.16.4]: "We shall explain the particular description of *bhū-gola* by name, form, measure, and characteristics." In the *Jyotiḥśāstra* the word *bhū-gola* refers to the earth as an egg of clay, and the word *viśeṣa*, or "particular," refers to the round golden egg described in the *Purāṇas*.

It is stated in the Second Canto, the *yoginīryāna* [SB 2.2.28], "then reaching the particular, or *viśeṣa*, one becomes fearless." It is also said in the Fifth Canto [5.20.35], "There is another land, made of gold, with a mirror-like surface," etc. Also, in the Third Canto [3.26.52] it is stated, "This universal egg is called particular or manifest (*viśeṣa*), with ten-fold increasing coverings." In the Fifth Canto [5.25.2]: "This great universe [*kṣiti-maṇḍalam* or earth-*maṇḍala*], situated on one of Lord Anantadeva's thousands of hoods, appears just like a white mustard seed." Thus, by the illustration of a mustard seed it is known to be spherical. It is also said in the *Kardama-vihāra* [3.23.43], "After showing his wife the globe [*golam*] of the universe [*bhuvaḥ*] and its different arrangements, full of many wonders, the great *yogī* Kardama Muni returned to his own hermitage." In the Tenth Canto [10.8.37] it is stated, "She [Yaśodā] saw within His mouth all moving and non-moving entities, outer space, and all directions, along with mountains, islands, the surface of the earth [*bhū-golam*], the blowing wind, fire, the moon and the stars." By such proofs one should accept that there are two earths.

Well, this earth is the smaller one, so where is the other, bigger one? The answer is: The bigger one is indeed a form of reflection [*pratibimba-rūpa*] up above the orbit of the asterisms. Its measure, according to SB 5.21.19, is "the 95,100,000-*yojana* circumference of the sun's orbit around Bhū." Thus roughly speaking, it comes to be the upper portion of the orbit of the asterisms.

Then how does the reflection appear, and how does it have this form? It is like this: On all sides of the earth of 5,000-*yojanas* circumference, separated at a distance of one *yojana*, there is the fire-sphere [*anala-golaḥ*]. Thereby, above the orbit of asterisms that seems small in the distance, the golden land of pure form creates a screen of light. Therein, on all sides, is the great reflection.

From a distant place a big thing looks small, and from another place a small thing looks big. Likewise, external objects of the universe such as the moon, look small, whereas the earth-globe, which is close by, looks big.

What is the evidence for the existence of the fire-sphere, or *anala-gola*? The evidence is that from the surface of the earth up to the limit of the orbits of the planets, there are eight divisions of winds, beginning with *āvaha,* and at the conjunction of the two [namely the earth and the winds], there is the water-sphere, or *jala-golaḥ*. It is mentioned in the *Puliṣa-siddhānta* that "the glass-like water-sphere is at the conjunction of the earth-air [*bhū-vāta*] and the *udvaha* wind, and by it the rays of the sun and other luminaries are seen separated and joined together.

According to *Siddhānta-śiromaṇi* [4.1–3 on p. 127], the following 7 winds are listed: *āvaha* or atmosphere, *pravāha*, *udvaha*, *saṃvaha*, *suvaha*, *parivaha*, and *parāvaha*. The atmosphere is 12 *yojanas* thick. The *pravāha* wind includes the fixed stars and planets, and sweeps them westward uniformly. Thus, the *jala-golaḥ* must be above the stars and planets.

The *Sūrya-siddhānta* [12.46] says, "Owing to closeness the sun's rays are vehement in summer [in the *devas*' regions]." Here "closeness" and "farness" could not exist without the reflection of the fire-sphere.

In accordance with the *Purāṇas*, Śukadeva Gosvāmī has spoken of the big measure of the earth, and only suggested the small measure; thus is the contradiction avoided by some. In the same way, although there is a contradiction involving the sphere of the sky, it is removed.

It is said in SB 5.21.2, "by this estimation [*etena*] the measurement of the circle of the *devas* is taught by those who are learned ..." The meaning is: *etena*, or "this estimation" means from the center of the bhū-globe to the circle [*maṇḍala*] of the *devas*, and it means the heights [*auccya*] and circumferences of the orbits of the planets and asterisms, etc.

The rest of SB 5.21.2 is: "just as the two halves of a grain of wheat meet in an intervening space, so the upper and lower halves of the universe meet in the region called *antarikṣa*, or outer space." This means that without measurement of the sphere of the earth and the *maṇḍala* of heaven there can be no measurement of *antarikṣa*, or outer space, just as there is no measurement of the grain between the two parts.

SB 5.21.3 states: "In the midst of the middle region [*antarikṣa*] is the most opulent sun." This means that the water-sphere is the seeming middle of *antarikṣa*. Just as *antarikṣa* lacks a center, so also the water-sphere lacks a center (middle) due to its sphericity. Thus the sun, which goes there, is established with the form of a reflection [*pratibimbha-rūpena*]. But the real sun disk [*bimba-rūpena*] is within 125,000 *yojanas* of the center of the earth.

SB 5.21.7 says [in paraphrase]: "The learned say the circumference of Mānasottara Mountain is 95,100,000 *yojanas*." The meaning is: *mānasas* means "the moon." *Uttaraḥ* means "others beyond the moon, up to Saturn." In accordance with *Jyotiḥśāstra*, the measure of its orbit, combined with the part of Saturn, comes to 127,800,000 *yojanas*.

According to SB 5.20.43, "The distance between the sun in the middle and the circumference of the universe [*aṇḍagola*, or "egg-sphere"] is 250 million *yojanas*." The meaning is: "Of the sun" means "of the reflected sun." "Egg-sphere" means "the circumference of the golden egg." The distance between the two would be 250 million *yojanas*.

SB 5.20.43 also says, "The sun is situated [vertically] in the middle of the universe, in the area between *dyaus* and *bhūmi* [Bhuvarloka and Bhūrloka, or heaven and earth] which is called *antarikṣa*, outer space." The meaning of this is: *Dyāv-ābhūmyoḥ* refers to the orbit of the asterisms and the earth. *Yad antaram* means "in the middle of the egg (i.e., the golden egg)." The sun therein is the sun of the water-sphere reflection. Thus it should be understood.

Such is the solution of the problem in the *Goladarpaṇa*; one should follow it. Thereby it is undoubtedly indisputable that by meaningful justification as to truth, the sacred *Bhāgavata*, being the statement of the supreme *āpta* [authority], is a means of proof, unrefuted, being in conformity with the conclusions of all sciences. The removal of contradiction in it, which is formerly written by me, that itself appears good. Why say more?

b. Description of the lineage of the commentator

By Śrī Vaṁśīdhara
English Translation by Krishna Swarūpa dāsa

These verses are taken from the concluding section of Śrī Vaṁśīdhara's *Bhāvārtha-dīpikā-prakāśa.*

|1|

bhuvana-pālana-tatpara-mānasād
giri-varāj jalapasya diśi sthitam |
hara-vilocana-yojana-bhūmike
raghu-sutasya nṛpasya sarovaram

"With the intention of nourishing or protecting the world exists the great mountain, Himālaya. To its western direction, within the area of about 3 *yojanas*, there is a water-pond (*sarovara)* which is named after Lord Rāma (Raghu-suta*)*."[1]

|2|

nikhila-tāpa-haram plavanādinā
tad upa-kaṇṭha-gataṁ śubha-pattanam |
kharaḍa-nāmakam asti janair yutam
śruti-vidhāna-parāyaṇa-mānasaiḥ

The water clears up all the struggles of everyone by flowing and flooding, etc. Close to it is a beautiful city name Kharaḍa where people follow scriptural regulations properly.

|3|

tasmin pure vipra gaṇāgra-gaṇya
uvāsa vipro balarāma śarmā |
tad atma-jo bhū-dhara-nāma-dheyaḥ
svācāra-niṣṭho 'khila-śāstra-vettā

In that city lived a leader of the *brāhmaṇas* named Balarāma Śarmā. His son was Bhūdhara (Śarmā) who knew all the scriptural injunctions and was fixed in his personal conduct.

|4|

gaurīśa-pādāmbuja-datta-cetā
gaurīprasādas tu tato babhūva |
tat sūnur audārya-guṇena yukto
jātaḥ sukarmā sukhadeva śarmā

Since he was fully dedicated to the lotus feet of Lord Śiva, his son was known as Gaurī Prasāda (Śarmā). His son was Śukhadeva Śarmā who became endowed with utmost liberal qualities.

|5|

tad deha-jo nigama-vid gajarāja śarmā
yo hy anvavāya-śubha-dīpaka āvir āsīt |
dharmātmakam madhuripor anugam sa nikvā-
rāmaṁ hy avāpa tanayaṁ vinayopapannam

His son Gajarāja Śarmā was born with beautiful qualities such that the whole dynasty became illuminated and was very well versed in all the Vedic literatures. His son Nikvārām became very religious, submissive, and a devotee of Lord Kṛṣṇa.

|6|

tasmāt vaṁśīdharo jāto gotre kauśiā-samjñake |
gauḍa-vaṁśe nanda-sūnu-pāda-dhyāna-parāyaṇaḥ

Vaṁśīdhara was born of that Gauḍa *brāhmaṇa* in the line of Kauśika-gotra as a son of Nikvārāma. He was constantly engaged in meditating and worshiping the lotus feet of Lord Kṛṣṇa, the son of Mahārāja Nanda.

|7|

tenāyaṁ grathitaḥ kṛṣṇa-kṛpayā grantha-sāgaraḥ |
śrīdhara-svāmi-bhāvārtha-dīpikāyāḥ prakāśakaḥ

By the grace of Lord Kṛṣna, he has compiled this ocean of literature which further illuminates the *Bhāvārtha-dīpikā*, the commentary (on *Śrīmad-Bhāgavatam*) by Śrīdhara Svāmī.

|8|

vyākhyā puṣpa-mayī mālā śrī vaṁśīdhara śarmaṇā |
samarpitā kṛṣṇa-kaṇṭhe prītyopādehī te mayā

This beautiful flower garland of explanation on [Śrīdhara Svāmī's] commentary is made by me, (Śrī Vaṁśīdhara Śarmā) and lovingly offered around the neck of Lord Kṛṣṇa.

|9|

śrī kṛṣṇa nātha pari-pūraya cintitaṁ śrī-
rādhā-pate na kaṭhinaṁ tava kiñcid asti |
tad dhyāna-niṣṭha-manaso mama yātu kālo
vṛndāvaneśa vraja-maṇḍala-gasya nityam

O Kṛṣṇa! O Lord! Nothing is impossible for you. Please fulfill my wish to reside within Vraja-maṇḍala, O beloved of Rādhā! O lord of Vṛndāvana! May my lifetime be spent being fixed always in meditating upon You.

|10|

śrī vṛndāvana-candrasya vṛṣārka-tanayā-pateḥ |
prīty arthaṁ ca kṛtaṁ granthaṁ śṛṇotu śrī hariḥ svayam

This book has been compiled for the pleasure of Śrī Vṛndāvana-candra, the beloved of the daughter of Vṛṣabhānu. May the Lord Himself kindly listen to it.

|11|

vṛṣabhānu-sutā-natha mām āhūya kṛtim mama |
vṛndāvane śṛṇu vibho hṛdayeśa nija-priya

O Lord of Rādhā, master of my heart and dear to your devotees, please call me in Vṛndāvana and listen to my work.

|12|

yad atra sauṣṭhavaṁ granthe tad guror eva me na hi |
yad atrāsauṣṭhavaṁ kiñcit tan mamaiva guror na hi

If there is anything acceptable in this book, that is due to the grace of my spiritual master. And anything unacceptable is coming from me, due to my shortcomings.

|13|

sarvāṇi prāktanāny eva svara-vākyākṣarāṇi vai |
kartṛtvaṁ me 'sti granthe 'smin teṣām yojana-mātrataḥ

I did not create anything new here. All the words, letters, and tones have already been existing. I'm the author of this book in the sense that I have simply used or organized the same words and letters in a certain way.

|14|

pramāṇī-bhūta-vākyāni pūrvācārya-kṛtāny api |
nija-grantha-prakāśena pūrvācāryā mati-pradāḥ

The previous *ācāryas* have bestowed intelligence upon me by publishing their commentaries with all the authoritative statements.

|15|

santi me guravaḥ sarve guravaḥ pitaro matāḥ |
pitrarjita-sve putrasya yathā svatvaṁ tathaiva hi

I have many gurus, the spiritual masters, and they are understood to be as good as fathers. So, just as the son gains ownership of his father's property, the same goes with the disciple.

|16|

teṣāṁ kṛtau sva-kartṛtvābhimāno viduṣāṁ varāḥ |
anyathā pitur artheṣu putrasyāpi na tad bhavet

This is the blessing of those previous scholars that I am holding the ownership of their work. Otherwise, the son would have no right to claim the ownership of his father's estate.

|17|

śrī kṛṣṇa-kṛpayā ceyaṁ prasṛtāskhaliteṣu ca |
vijñeṣu te hi cāpy atra nyūnatāṁ pūrayantu vai

By the grace of Lord Kṛṣṇa, this book has expanded and is firm amongst the scholars. Still, I request them to kindly make corrections if there are any mistakes here.

|18|

mayy alpa-jñe kṛpāṁ kṛtvā nija-sauṣṭhava-bhāvataḥ |
bhavantv evarju-tatva-jñe paṇḍitāḥ sudayodayāḥ

As I'm so ignorant, may the wise ones being very merciful upon me due to their own great qualities, become my guide to understand the truth in a right manner.

Date of Completion

|19|

śarāmnāyāṅka-śītāṁśu-mite samvatsarorjake |
site viṣṇu-tithau pūrṇo vṛndāraka-guror dine

This book was completed in the year 1945 of Vikrama Saṁvatsara [1888 CE], on Ekādaśī (eleventh day) of the bright fortnight of month of Mārgaśīrṣa, on Thursday.[2]

Meaning of words in the system of astrological language which goes right to left:

Śara refers to the arrows of Kāmadeva, that are 5.
Āmnāya indicates the Vedas, which are 4.
Aṅka refers to total numerical digits, which are 9.
Śītāṁśu means the moon which represents number 1.
Saṁvatsara means the year, that is 1945 [1888 CE], as mentioned above.

Ūrjake means in the month of Mārgaśīrṣa, right after the Kārtik.
Site means during the white fortnight.
Viṣṇu-tithau refers to the eleventh lunar day, on Ekādaśī.
Pūrṇaḥ means it was completed.
Vṛndāraka means the best.
Guror dine means on the day of Jupiter, Thursday.

c. A historiographical account of biographical material on the life of the Vaishnava commentator, Śrī Va□śīdhara

by S. E. Kreitzer

Śrī Vaṁśīdhara is recognized within a legacy of respected Vaishnava commentators, some of whom considered the apparent discrepancies suggested by cosmographical descriptions found within Purāṇic literature when compared with the Sanskrit mathematics of observable astronomy found within the *jyotiṣa* tradition. Both were considered respected traditions integral to Vedic culture. Śrī Vaṁśīdhara's commentary on the *Śrīmad-Bhāgavatam*, addressing these topics, suggests a devotionally insightful analysis grounded in Siddhāntic integrity.

As part of the research for his pioneering work, *Vedic Cosmography and Astronomy* (1989), Richard L. Thompson secured a complete translation of Śrī Vaṁśīdhara's *Śrīmad-Bhāgavatam* 5.20.38 purport from two well-regarded Sanskrit scholars, Dr. V. N. Jha from the University of Pune and Dr. V. V. Gangal from the Ananthacharya Indological Research Institute. Śrī Vaṁśīdhara's comprehensive Sanskrit commentary on the *Bhāgavatam* intentionally built upon Śrīdhara Svāmī's foundational exegesis written centuries earlier. In a sense, Vaṁśīdhara represents a bookend to Śrīdhara within an esteemed tradition of Vaishnava scholarship, as his work would be one of the last major traditional commentaries written entirely in Sanskrit. As the 19th century progressed, the tradition shifted noticeably toward works written in vernacular languages such as Bengali, Hindi, and English, and since then, in a variety of languages spoken throughout the world.

Whereas Thompson and other scholars had been considering Vaṁśīdhara as flourishing during the 17th century, they admitted to having limited access to historical evidence to do so. More recent research leans toward the 19th century, placing Vaṁśīdhara as a near contemporary of Bhaktivinoda Ṭhākura. Support for this latter date is featured in a short biographical sketch written by Dr. Brahmanand Tripathi that drew from references in Vaṁśīdhara's work. Tripathi would also include a late 19th-century photograph as part of the front matter to a reprint of Śrī Vaṁśīdhara Śarman's Sanskrit *Bhāgavatam* commentary, *Bhavārtha-dīpikā-prakāsa*, which literally translates as "Illumination of Śrīdhara Svāmin's Lamp on Sense and Meaning." Chaukhamba

Sanskrit Pratishthan reprinted this work in 1993 and then again in 2013, and marketed it with an English title, *Srimad Bhagavata Purana with Two Commentaries – Srihari and Vanshidhari* (Set of 2 Volumes). Tripathi's account of Vaṁśīdhara Śarman flourishing during the 19th century offers a strong argument.

As for scholars considering an earlier date, in his 1954 essay published in the *Journal of the Oriental Institute*, Parashuram Krishna Gode considered the possibility of Vaṁśīdhara writing during the 17th century: "In the genealogy of Navanītarāma given by him we find *Vaṁśīdhara*, son of *Bhāgavanla*. Is he identical with Vaṁśīdhara, author of the *Bhāvāthadīpīka* commentary … ?" Gode identified the work associated with Vaṁśīdhara as "published by the Venkatesvara Press, Bombay,"[3] in a five-volume set published in 1907 (sometime referenced as 1908). Gode added, "If this identity is correct," then Vaṁśīdhara's life dates would place him after Śrīdhara, and thus "*later than c. AD 1450*," while prior to "the earlier terminus for the date of Navanītarāma *viz*. c. AD 1650."[4]

In *Vedic Cosmography and Astronomy*, Richard Thompson located Vaṁśīdhara with a similar date, "living in AD 1642, before the time that Western science made a large impact on India."[5] Thompson identified the author's full name as Vaṁśīdhara Śarman in notes for an interactive digital presentation describing Fifth Canto cosmology. Coincidently, the same date used by Thompson is found in the Forward to *Prakrta-Paiṅgalaṁ: A text on Prakṛita and Apabhraṁśa Metres* Part I (1959), where V.S. Agarwal wrote about the "'Piṅgala-Prakāśa' of Vaṁśīdhara (1642 AD)."[6] The first edition of this set contains three Sanskrit commentaries, one identified with Vaṁśīdhara. Agarwal repeats this information in the second edition supplement (1962), on page 13.

Perhaps Thompson drew the 1642 date from Agarwal's account. In a presentation given at a 1990 Vrindavan Institute for Higher Education seminar held in Port Royal, Pennsylvania, Thompson noted that while "Vaṁśīdhara was active in 1642, I don't know any other dates concerning his life. But according to what I was able to find he was writing things, [and was thus] mature in 1642 AD"[7]

Thompson acknowledged that his primary exposure to Vaṁśīdhara's work was drawn from Kṛṣṇa Śaṅkara Śāstrī's nine-volume set of traditional *Bhāgavatam* commentaries, published beginning in 1965. Śrīla Prabhupāda respectfully acknowledged Śāstrī's scholarship in a letter written to Sumati Morarjee, August 9, 1976. Prabhupāda noted that Śāstrī's work *Maharsīvedavyāsapraṇītam Śṛīmadbhāgavatamahāpurāṇam*, had given "almost all the important commentaries on the *Bhāgavatam*." These commentators included "1. Sridhara Svami 2. Sri Vamsidhara," as well as, "6. Srimad Jiva Goswami 7. Visvanatha Cakravarti Thakura 8. Srimad Sukadeva," among others. Śrī Vaṁśīdhara Śarman's status as a respected Vaishnava commentator appears secure in this assemblage.

In a second example illustrating the significance Prabhupāda placed in Śāstrī's compilation – it appears he traveled with this volume as suggested by the following excerpt taken from a conversation held in Toronto (June 18, 1976). Among the participants in this exchange were Joseph T. O'Connell from the University of Toronto, a noted religion scholar specializing in Caitanya Vaishnavism, along with Bimal Krishna Matilal, Professor of Sanskrit at Toronto who the following year would be elected Spalding Chair of Eastern Religions and Ethics at Oxford University, and Professor Shivaram from McMaster University.

> **Guest:** Is there any other *bhāṣya* other than *Baladeva-bhāṣya*?
>
> **Prabhupāda:** No, many others.
>
> **Guest:** Gauḍīya, Śaṅkara ...
>
> **Prabhupāda:** No, other *ācāryas*. Rāmānujācārya, Madhvācārya, then Vallabhācārya. Then our *Gauḍīya-bhāṣya*, this Jīva Gosvāmī, Viśvanātha Cakravartī Ṭhākura. And Rādhāramaṇa Gosvāmī, Gopāla Bhaṭṭa Gosvāmī's descendant.
>
> **Guest:** Is there any plan to translate into English these things? Is there any plan to translate Jīva Gosvāmī's, *Jīva-bhāṣya*?
>
> **Prabhupāda:** No, no. I am taking help from all these Gosvāmīs and giving a summary. Where is that book? That green book? (referring to Prabhupāda's green *Bhāgavatam* with eight commentaries) (break) First comment I find, Viśvanātha Cakravartī and Vīra-rāghavācārya, the Rāmānujācārya Sampradāya. Vīra-rāghavācārya.... So

read the names of the commentator, any chapter. Who are ... Śrīdhara Svāmī.

Pradyumna: (reading Sanskrit names) Śrīdhara Svāmī, and Vaṁśidhara *kṛta Bhāvārtha-dīpikā-prakāśa bhāṣya.* Śrī Gaṅgā Sahaya *praṇītā Anvita-prakāśa.* And Śrīmad Vīra Rāghavācārya, *Bhāgavata-candrikā.* Śrīmad Vijayadhvaja Tīrtha *racita Pada-ratnāvalī.* Śrīmad Jīva Gosvāmī, *Krama-sandarbha.* Viśvanātha Cakravartī *kṛta Sārārtha-darśini.* Śrīmad Śukadeva *kṛta Siddhānta pathī.* Gosvāmī Śrī Giridhari Lāla Gupta, *Gośāla-prabodhinī.*

Prabhupāda: That is Vallabhācārya.

Pradyumna: ... Śrī Bhagavatprasād Ācārya *praṇīta Bhakti-mano-rañjanī.*

Prabhupāda: All the *ācāryas*, their disciplic succession, they have written their comments.[8]

Another noted scholar who has considered Śrī Vaṁśidhara living during the 17th century, is the American Indologist from the University of Washington, Karl Potter, who edits the multi-volume *Encyclopedia of Indian Philosophies.* In Volume XIX subtitled *Acintyabhedhābheda Vaiṣṇava Philosophy*, he offers a description of Vaṁśīdhara that acknowledges his "commentary on the *Bhāgavatapurāṇa*," as well as a life date of "(1680?)"[9] In an online edition of the *Bibliography of Indian Philosophies*, updated April 15, 2020 (1st edition 1970), Potter similarly identifies Vaṁśīdhara with the reference "1268A. Vaṁśīdhara (1680?)"[10] Like Thompson, Potter acknowledges Śāstrī's nine-volume set of *Bhāgavatam* commentaries as a reference for Vaṁśīdhara's work.

More recent life dates are drawn from what appears to be a facsimile reprint of the 1907 Sanskrit edition of Vaṁśīdhara's *Bhāvārtha-dīpikā-prakāśa* produced by Sri Venkatesvara Press, as previously mentioned in P. K. Gode's 1954 essay. As noted, reprints published in 1993 and 2013 by Chaukhamba Sanskrit Pratishthan include front matter by Dr. Brahmanand Tripathi, whose name is associated with a variety of traditional Sanskrit and Hindi language texts published during the latter

20th century. Tripathi noted toward the end of his forward, written in Sanskrit, that Vaṁśīdhara had mentioned in the concluding portion of his Sanskrit commentary his personal background as a Gauḍa Brahmin born in the line of the Kauśika Gotra identified with Viśvāmitra Muni. Vaṁśīdhara described living in the Himalayan region, and also referred to a district known as Karada. While it is not entirely clear exactly the area "Karada" refers to, contemporary maps locate a Karada village in Rajasthan. The name Śrī Khaḍa is also found in scholarly translations of Vaṁśīdhara's work as a reference to a locale where he had spent time writing.[11]

Vaṁśīdhara would secure patronage from Sir Hira Singh Malvendra Bahadur, Maharaja of Nabha, a princely state in southeast Punjab about 55 miles (90 km) southwest of Chandigarh, which is located by the Himalayan foothills. Hira Singh (1843–1911), during his forty-year reign, earned a reputation as a competent and pious administrator who not only supported traditional Sanskrit scholarship, but instituted modernization programs involving infrastructure, agriculture, healthcare, and education. The Maharaja also maintained an efficient army utilized in support of the British Raj.

As for Vaṁśīdhara, Tripathi wrote that he lived approximately between 1828 and 1890. He noted that Vaṁśīdhara had initially published his work with the Veṅkaṭeśvara Press in the Vikram year 1945 (1888 CE). Tripathi described that more information is offered by the commentator in the concluding portion of his Sanskrit commentaries.[12]

Jonathan Edelmann, Assistant Professor in the Department of Religion, University of Florida, offered an analysis of Vaṁśīdhara's commentary in the *Journal of Vaishnava Studies*, in an essay titled "Vamshi Dhara Sharman's *Bhāvârthadīpikāprakāśa* and the Original-Reflection Theory." According to Edelmann, "Vamshi's specific contribution to commentaries on the *Bhāgavatapurāṇa*," describe that the "highest reality transcends human language and requires a re-organization of our words and concepts to comprehend," and further, "the experience (*upalabhyamāna*) of this ... is real and the foundation for how we read and interpret scripture."[13] Titled "Illumination of Śrīdhara Svāmin's Lamp on Sense and Meaning," Vaṁśīdhara's commentary "often quotes in an affirmative manner Gauḍīya Vaishnava theologians like Jīva Gosvāmin and Viśvanātha Cakravartin."[14] Additionally, it "echoes that of Jīva Gosvāmin, Viśvanātha, and other Vaishnava commentators like Madhva and Vijayadhvaja."[15]

As for biographical perspectives on Vaṁśīdhara, Edelmann writes:

> He might have lived in Mathura, Uttar Pradesh, in North India, at some point, as suggested here: "By the urging of the learned people of Mathurā, I, Shri Vamshi Dhara Sharman, for the satisfaction of Shri Krishna, who descended into the circle of Mathura, have undertaken to begin a commentary on Śrīdhara's "A Lamp of Sense and Meaning." His colophon on Book 12 says he composed his commentary while living among the people of Śrī Khaḍa, but it is not clear if he wrote all of his commentary there or just a part of it.[16]

Edelmann adds that "Vamshi's *Bhāva-artha-dīpikā-prakāśa* was likely 'composed in 1888–89,'" with the citation referencing H. G. Shastri's edited volume, *The Bhāgavat [Śrīmad Bhāgavata Mahapurāṇa]*.[17] Edelmann further reports that "the *Bhāvārthadīpikāprakāśa* is the only book listed under Vamshi Dhara's name in the *New Catalogus Catalogorum*," and that his own source for studying Vaṁśīdhara's work is "primarily confined to Kṛṣṇa Śaṅkara Śāstrī's monumental publication, the *Śrīmad Bhāgavata Purāṇa*"[18] – the same volume referenced by Śrila Prabhupāda in his previously mentioned letter to Sumati Morarji.

Concerning K. S. Śāstrī's source material for publishing Vaṁśīdhara's *Bhāvārtha-dīpikā-prakāśa*, Edelmann writes:

> I believe that Śāstrī's edition is based on a lithographic edition of the text published in Mumbai in the early twentieth century by Śrīveṅkateśvara. This edition is currently held by the Leiden University Library in the Netherlands and was recently republished in the Chaukhamba Sanskrit Pratishthan in New Delhi.[19]

The catalog reference offered by the Leiden University Library describes the physical work as being in loose leaves. It appears to also note 1867 as an earlier publication date.

As for other references of note pertaining to Vaṁśīdhara:

(1) *The Supplemental Catalogue of Sanskrit, Pali, and Prakrit Books in the Library of the British Museum acquired during the years 1892–1906*, compiled by L. D. Barnett and published in 1908, offers two references to the following work (Devanāgarī script not included) ascribed to Vaṁśīdhara:

> 14093. b. 44. (page 618) – [Sārasvatasūtra. With the commentary of Laghu-bhāshya of Raghunātha son of Vināyaka, and as an appendix the Sūtra alone. Edited by Vaṃṣīdhara of Nabha.]. pp. ii. 460. [*Bombay*, 1901.] 8°.

> 14093. b. 44. (page 781) – Vaṃṣīdhara, *of Nabha*. See Sārasvatasūtra. [Sārasvatasūtra. With the Laghubhāshya of Raghunātha. Edited by Vaṃṣīdhara.) [1901.] 8°.

(2) There is another reference in the second volume of the *Library of the British Museum* supplemental catalog. This volume offers works "acquired during the years 1906–1928":

> 14101. c. 1. (page 253) – [Pramāṇa-naya-tattvālokālaṅkāra. With Ratnaprabha's commentary Ratnākarāvatārikā, and a Hindi paraphrase of the whole by Vaṃṣīdhara Ṣarmā.] ff. iv. 157. [*Unja*, *Bombay* printed, 1910.]. obl. 4°.

(3) On page 149 of *A History of Indian Literature: The Purāṇas*, Ludo Rocher discussed the reputation of Śrīdhara Svāmī's *Bhāvārtha-dīpikā* amongst the Bengali Vaishnavas. Rocher noted that alongside eminent commentators such as Jīva Gosvāmī and Viśvanātha Cakravartī, "Two other Bengal Vaiṣṇavas wrote commentaries on the Bhāvārthadīpikā: Rādhāramaṇa Gosvāmin (-dīpanī), and Vaṁśīdhara Śarman (-prakāśa)."[20]

(4) The frontmatter to the Sanskrit volume, *Śrīmad-bhāgavata-mahāpurāṇam* with the *Gūḍhārtha-dīpinī* Sanskrit Commentary, Volume 1, Skandhas 1–9, offers images of Śrīdhara Svāmī and Vaṁśīdhara, along with photographs and contextual information for a number of

sādhus and *ācāryas* also identified with the publication. The image of Vaṁśīdhara is a slightly cropped version of the photograph included in Tripathi's 1993 Chaukhamba Sanskrit Pratishthan edition.[21]

(5) In Edwin Bryant and Maria Ekstrandt's edited compilation, *The Hare Krishna Movement: The Postcharismatic Fate of a Religious Transplant*, the footnotes to Chapter 21 written by Ekkehardt Lorenz include life dates for Vaṁśīdhara Paṇḍita as a *Bhāgavatam* commentator.[22] In an email exchange with Prishni Sutton (February 6, 2021), Lorenz references his source to Dr. Brahmanad Tripathi's previously mentioned volume of Vaṁśīdhara's commentary published in 1993 by Chaukhamba Sanskrit Pratishthan. Lorenz acknowledged that upon closer examination: "I noticed that I gave the date as 1820–1890 (in my footnote 18 on page 126), while Brahmand Tripathi actually wrote: '*anenānumīyate yad amuṣya sthiti-kālaḥ* 1828 *taḥ* 1890 *khrīṣṭābda-madhye samabhavad iti*.' "

APPENDIX 5

Evidence for a Spherical Earth

by Antardwīp dāsa

Section 1: Mathematical Proof Based on *Sūrya-siddhānta* 12.38–41

"At a quadrant of the earth's equator in the east, in Bhadrāśva-varsa, is the city famed as Yamakoṭi, having walls and gateways of gold. To the south, in Bhārata-varsa, is in like manner the great city Lankā. In the west, in Ketumāla-varsa, is declared to be the city named Romaka. In the north, in Kuru-varsa, is declared to be the city called Siddhapurī; in it dwell the perfected great souls, free from trouble. These are situated at equal distances from one another on a quadrant of the earth's equator. To the north of them, at the same distance, is Meru, the abode of the gods."[1]

Proof

The four cities are at quadrants from each other on the earth's equator. That would be the same whether the layout was on a globe or a flat, round earth. However, when Meru is presented as being at the same distance from them, then the flat earth is ruled out. Let me explain.

The equator is a circle, hence a quarter of it is a quadrant, or 90 degrees. If the earth were flat and circular, then the distance from the North Pole in the center to the equator as a circle would be a certain amount, let us call it r (for radius); it doesn't actually matter how much it really is in kilometers, miles or *yojanas*. The equator, being a circle, will therefore have a circumference of $2\pi r$. A quadrant of that, or a quarter, will therefore have the length of $\pi r/2$. In comparison to that, note that the distance from any of these four cities to the North Pole will be exactly r by definition.

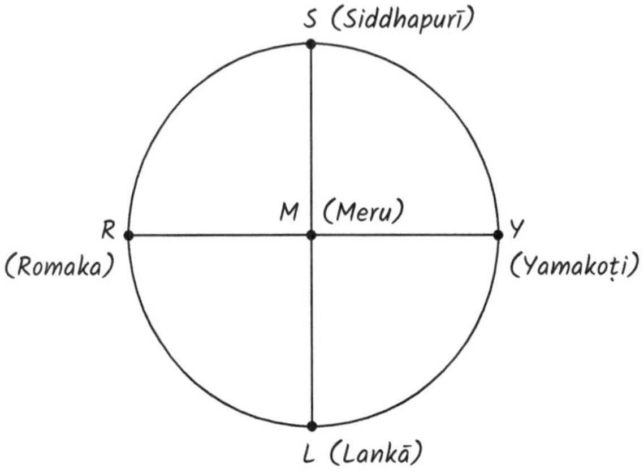

Figure 32: *Sūrya-siddhānta* 12.38–41 mentions four cities (Yamakoṭi, Lankā, Romaka, and Siddhapurī) at an equal distant from each other on each quarter of the earth's equator, represented in this image by the circle's circumference.

Obviously r does not equal $\pi r/2$. Therefore the layout of a flat and round earth does not match the description given in the *Sūrya-siddhānta*.

Now, let us consider the earth as a globe. The equator is a circle, again we will give it a radius of r, and each of the cities located at a quadrant upon it is also at a distance from the next of $\pi r/2$. This needs no further explanation.

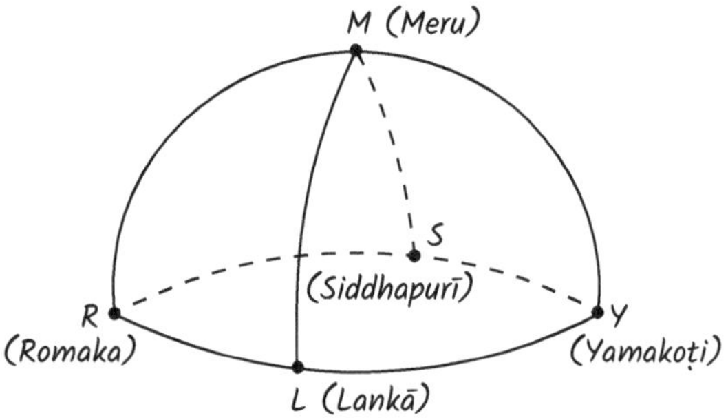

Figure 33: The four cities at quadrants from each other on a line diagram of the Northern Hemisphere of a spherical earth.

Importantly, any circle running through both the North and South Poles, which will thus be composed of two meridians located 180 degrees apart, is a circle the same size as that of the equator. This is necessarily the case as both types of circles are great circles on the sphere, and by definition all great circles have the same circumference and same radius.

Note that the distance from any of these equatorial cities to the North Pole will therefore be one quarter of a great circle, which has the same radius as the equator. Therefore, the distance is $\pi r/2$, the same as that between two adjacent cities. Hence, the globular form of the earth matches the description given in the verses. QED

The following translation of *Sūrya-siddhānta* 12.38–41 is from the *Sūrya-siddhānta and Siddhānta-śiromaṇi* (*Graha ganita'dhyaya*), compiled and edited by Danavir Goswami (2007):

> East of the fourth quadrant of the earth-globe is *bhadrāśwa-varsha* and in the middle of that is *yamakotipuri*. It is said that this one has golden walls and golden gateways. To the south, is Bhāratavarsha, in the midst of which is the great city Laṅkā. In the west, in Ketumāla-varsha is the city named Romaka. In the north, in Kuru-varsha, the perfected great souls live free from all agonies.
>
> These cities are situated at an equal distance from each other on each quarter of earth's equator. At an equal distance from them, to the north is situated the Uttara Meru, the abode of the demigods.

Section 2: StarTrails

A star trail photograph uses long exposure photography to capture the motion of stars at night, as viewed from the surface of the earth. In a star-trail photograph, individual stars appear as streaks across the image.

Figure 34: Photograph of star trails viewed from Perth, Australia, taken the the author.

This image was taken on Saturday night, March 13, 2021, between the hours of 10 pm and 1 am, in a forest that is about half an hour east of Perth, Australia. The location is called Forsyths Mill, and the exact camera set-up spot was 31° 53' 31.7" South, 116° 16' 44.1" East. The camera was pointing South.

APPENDIX 6

Polar Projection Maps and Flat Earth Arguments

By Prishni Sutton

Contemporary flat earth advocates endorse a number of maps they claim to literally represent a flat earth. The model frequently put forward is the equivalent of a northern polar equidistant azimuthal projection map (Figure 35).

These maps are created by placing a flat plane tangent to the North Pole and then projecting a line from the South Pole through any point on the globe to where it intersects this plane. The most widely known representation of an azimuthal equidistant projection map is the image of the earth on the United Nations flag (Figure 36).

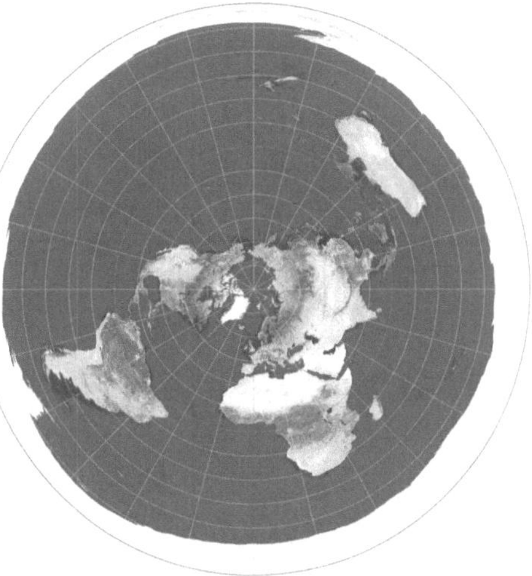

Figure 35: An image of a northern polar equidistant azimuthal projection map. (Image by Daniel R. Strebe.)

Figure 36: Flag of the United Nations displaying a northern polar equidistant azimuthal projection map of the globe.

Suffice it to say, the overwhelming majority of UN members appear to consider the earth a globe while viewing the UN flag. This section will examine a variety of polar projection images, both historical and recent, that present two-dimensional maps of the geographically spherical earth.

Flat circular world maps have been used to represent a spherical earth for centuries. The renowned cartographer Gerardus Mercator's world polar projection map published in *Atlantis Pars Altera* (1595, with features represented earlier in his 1569 world map), offers a classic example. It also represents the first map of the Arctic region (Figure 37).

Figure 37: Gerardus Mercator's world polar projection map published in *Atlantis Pars Altera* (1595). For more information please see: https://library.princeton.edu/visual_materials/maps/websites/northwest-passage/mercator.htm

Along similar lines, Figure 38, from Andreas Cellarius Palatinus' *Harmonica Macrocosmica* (1661), featured an illustration of Ptolemy's *Planisphaerium* (c. 150 CE) in which the earth map (Figure 38)

at the centre of the planetary system is shown in great detail, with the North Pole in the centre, America in the upper half and Europe and Africa in the lower half. A silver chariot on a silver-white circle indicates the path of the Moon, a golden chariot on a golden circle the path of the Sun. The planets are found on the remaining circles.[1]

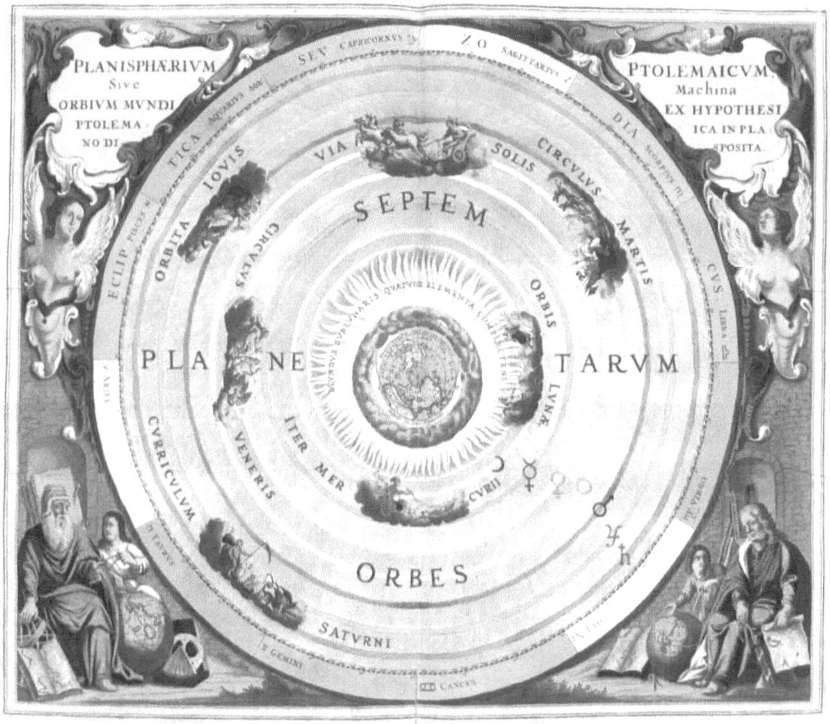

Figure 38: An illustration of Ptolemy's *Planisphaerium*, from Andreas Cellarius Palatinus' *Harmonica Macrocosmica* (1661). For more information please see: https://collections.leventhalmap.org/search/commonwealth:x633f9294.

The famed French cartographer and astronomer Jacques Cassini also crafted a beautiful northern polar projection map in 1713, which remains popular as wall art (Figure 39).

A recent two-sided map (Figure 40) designed by the Princeton astrophysicists Richard Gott and Robert Vanderbei, along with the physicist David Goldberg of Drexel University, preserves the functionality of earlier projection maps, while updating the geometric proportions experienced on the surface of a globe.[2]

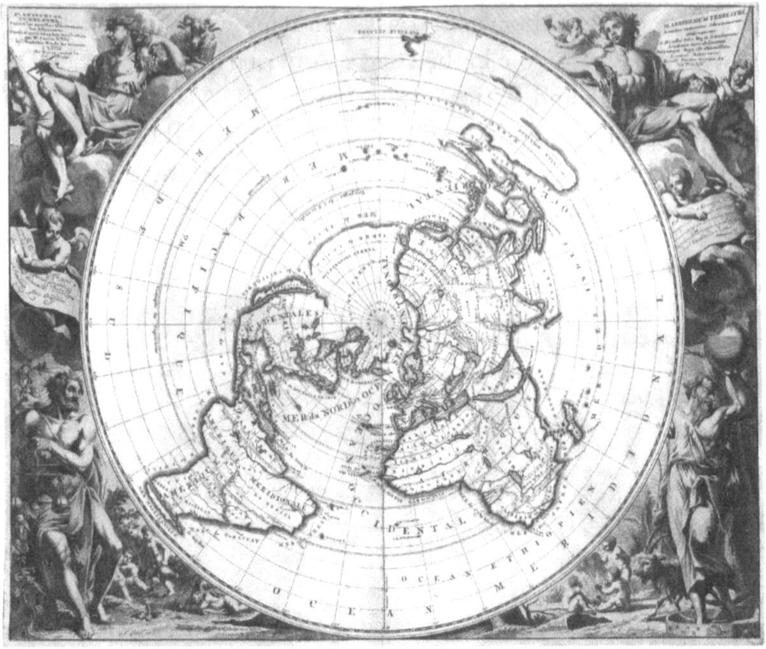

Figure 39: Northern polar projection map by Jacques Cassini with the title, *Planisphere terrestre. Suivant les nouvelles observations des astronomes* (1713). For more information please see: https://digitalcollections.lib.washington.edu/digital/collection/maps/id/13/.

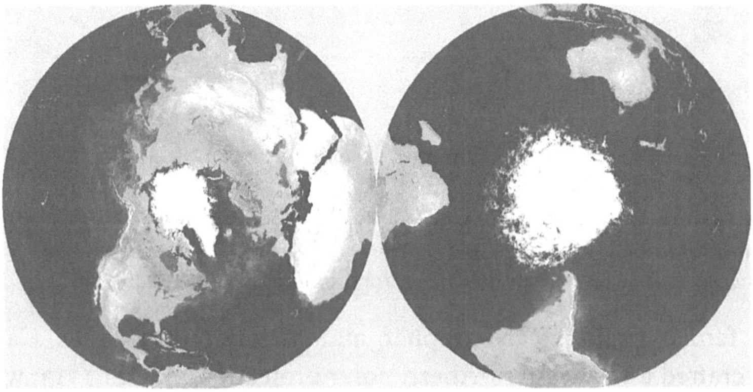

Figure 40: In 2021, J. Richard Gott, Robert Vanderbei, and David Goldberg developed a "revolutionary, double-sided map" that "minimizes all ... types of distortions." For more information please see: https://www.princeton.edu/news/2021/02/15/princeton-astrophysicists-re-imagine-world-map-designing-less-distorted-radically.

There are several well-known maps offered by flat earth advocates that closely resemble northern polar projection maps. For example, there is an oft-cited illustration from 1931 identified with the evangelist Wilbur Glenn Voliva (Figure 41), as well as an 1892 "New Standard Map of The World" credited to Alexander Gleason (Figure 42).

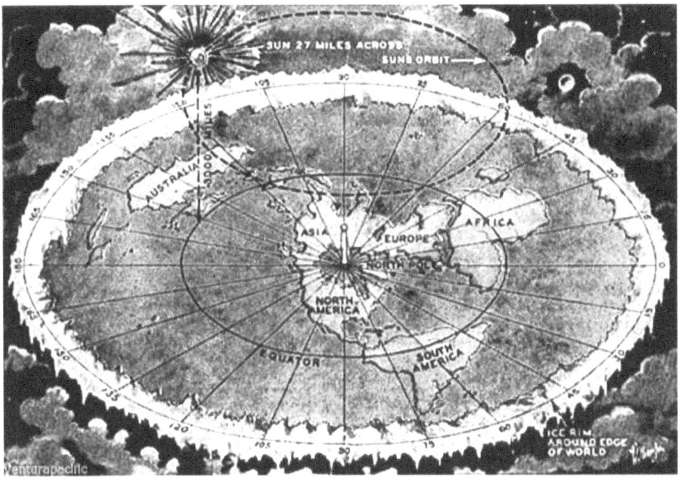

Figure 41: A map of Wilbur Glenn Voliva's "Flat Earth" published in *Modern Mechanics and Invention* (October 1931), in an article by James Earle Miller titled "$5,000 for Proving the Earth a Globe." Miller concluded that the prize money offered by Voliva to anyone who could convince him of a spherical earth, "probably will remain uncollected unless some future space traveler some day anchors his ship a few thousand miles out in space and takes a movie of a globular world turning on its axis."

Frequently lacking with images such as these are accompanying analyses comprehensively explaining how they could have been be created by methods other than those used for a classic polar projection map, which historically identified features of the globe in two dimensions.

Figure 42: Alexander Gleason's "New Standard Map of the World" (1892) representing ideas of the Flat Earth Society. Gleason is also credited with authoring the book, *Is the Bible from Heaven? Is the Earth a Globe?* (1890), offered for download at the following address: https://archive.org/details/alex_gleason-is_the_bible_from_heaven__is_the_earth_a_globe.o/

APPENDIX 7

Sanskrit Mathematics in Western Antiquity

By S. E. Kreitzer

Even well outside of India, the *jyotiṣa* tradition received high praise. During the late Roman period, not long after its greatest expanse from its eastern capital in Constantinople under the rule of Justinian the Great,[1] there lived the "noted 7th-century Syriac Orthodox polymath and teacher," the Christian bishop and philosopher Severus Sebokht (b. 575 CE), who knew well the mathematical achievements of the astral *paṇḍitas* of India. Coincidentally, Bishop Severus' branch of Christianity shares historical roots with the Saint Thomas Christians established in India shortly after the departure of Jesus of Nazareth.[2]

Known for his works on astronomy and geography, Severus is explicitly acknowledged for having "played an important role in the transmission of Indian intellectual concepts into Syria and ultimately the Islamic world." For example, in "one famous passage, he praises the Hindu decimal concept and mentions for the first time in the Greek east the nine numerical symbols used in India."[3] Around 662 CE, Severus shared this information in written exchanges with Basil of Cyprus, a fellow priest and teacher with whom he shared a number of "astronomical, mathematical and cosmographical questions."[4]

In the historical manuscripts that remain extant, Severus attempted to set the record straight concerning an apparent hubris noticeable amongst Greek scholars at the time. In good part to drive this point home, Severus discussed with profound respect the mathematical astronomers of India. For Severus, the *jyotiṣa paṇḍitas* illustrated the universal nature of genuine knowledge:

> I shall not now speak of the knowledge of the Hindus . . . of their subtle discoveries in the science of astronomy—discoveries even more ingenious than those of the Greeks and Babylonians—of their rational system of mathematics, or of their method of calculation which no words can praise strongly enough—I mean the system using nine symbols. If these things were known by the people who think that they alone have mastered the sciences because they speak Greek they will perhaps be convinced, though a little late in the day, that other folk, not only the Greeks, but men of a different tongue, know something as well as they.[5]

Historians have considered this segment written by the Bishop Severus Sebokht as the first known evidence in the West about a discussion of the Indian decimal system, more generally known as the Hindu-Arabic numeral system, since its eventual popularization outside of India drew from Islamic scholars.[6] A. L. Basham, the famed historian of Asian studies, wrote in his classic work, *The Wonder That Was India* (1954), of the transmission of the decimal system:

> The debt of the Western world to India in this respect cannot be overestimated. Most of the great discoveries and inventions of which Europe is so proud would have been impossible without a developed system of mathematics, and this in turn would have been impossible if Europe had been shackled by the unwieldy system of Roman numerals.[7]

Basham added to that:

> The success of Indian mathematics was mainly due to the fact that the Indians had a clear conception of abstract number, as distinct from numerical quantity of objects or spatial extension. While Greek mathematical science was largely based on mensuration and geometry, India transcended these conceptions quite early, and, with the aid of a simple numeral notation, devised a rudimentary algebra which allowed more complicated calculations than were

> possible to the Greeks, and led to the study of number for its own sake.[8]

Though Severus could criticize the pride of Greek scholars as beholders of ancient learning, he also discussed with Basil of Cyprus a summation of his thoughts:

> I have said these things not because I despise the wisdom of the Greeks in such matters ... but because I want to show that knowledge is the common property of anyone who wishes to be diligent, regardless of whether he is Greek or non-Greek ...[9]

Interesting to note how Severus wrote much the same as his South Asian contemporary, Varāhamihira, who pointed out that the central value of the work of the *brāhmaṇa jyotiṣa paṇḍitas*, or likely any pundit East or West, could be gleaned from their scholarly diligence. Without question, scholarly evidence drawn from ancient times suggests the high stature commanded by the Sanskrit mathematicians of the Indian subcontinent.

APPENDIX 8

Personal Testimony: "A View from 41,000 Feet"

A Conversation with Emmanuel Namamrta Greene

July 2, 2021

Prishni Sutton: Hare Krishna, Namamrita Prabhu. I'd like to discuss your experiences concerning the shape of the earth from your vantage point as a commercial pilot for Southwest Airlines. But first, please begin by telling us a little bit about yourself.

Namamrita: Alright. I was born at the Philadelphia temple in 1979. At that time my father, Yogesvara, was the temple president and eventually, when I was around two or three, we moved back to France where my mother, Jyotirmayi, is from. I grew up in France until I graduated from high school. That included living in temples, and attending *gurukula* through the age of 12, when I switched to public school. After graduating, I moved to New York to complete a four-year aeronautics degree, a BS in aeronautics.

P: How did you get interested in aeronautics?

N: My interest in flying could be from a previous life, who knows? But my earliest memories go from jumping off the jungle gym with an umbrella in my hand trying to mimic Mary Poppins, to laying face down in a *dhoti* harnessed at the end of a rope hanging off of a tree branch and running in concentric circles wider and wider, until eventually my feet would lift off the ground. I was flying! And every day I would look forward to going back to that tree and taking to the air once more. So, there's been something there ever since a young age.

P: Some propensity. How did it develop from there?

N: Yes. Eventually I developed a strong passion for remote control airplanes, and that's where I began learning about some of the science behind it. My very first remote control airplane was a two-meter glider, which took me a long time to build from a balsa wood kit, and I used to have dreams of finally flying it. At the RC airplane field I used to go to in the south of France, my first aviation mentor Jean Luc drew an airfoil in the gravel by our feet and explained some basic concepts such as how lift was created, and that fascinated me. When I moved to the U S and began my aeronautics degree, I had already understood many of the concepts and was able to learn the technical terms behind them and further my understanding of the fundamentals of flight and the different forces involved. I was excited to discover more about the physics of flight, and learn the language of aviation. I really enjoyed my college experience and loved the idea that I could make a living out of something that brought me so much joy.

P: So, when did all this happen?

N: While living in France I would visit my father in New York twice a year. This began at age six as an unaccompanied minor. I spent a whole lot of time on airplanes, in this way, and back in the pre-9/11 era they would often bring me into the cockpit for both takeoff and landing. I got to see a lot of aviation in this way, including 747s, and I remember how, as a young boy, I would study the movement of the different panels on the wings looking out the passenger windows and try to understand what it meant. In the summer of 1997, while in New York, my grandmother, Adele Greene, paid for me to have an introductory flight in a Cessna 172. She sat in the back, and takes full credit for my aviation career. When we got back from that flight my dad asked me, "So how did you like it?" I replied that I loved it and he said, "There's an aviation college right here in Long Island. If you're interested in making a career as a pilot, I'll set you up for it." My love for flying was so strong that I was willing to leave my mother and brother, French culture, and everything I knew behind to come and live in the U.S. and learn how to fly.

I moved to New York in 1998, enrolled at the Dowling College School of Aviation, and immediately began my aviation courses. The first courses I took were called "Basics of Flight" and "Aviation Weather." They took us through all the ratings, the private pilot license, the instrument license, the commercial license. I was getting college credit for all of those. I enjoyed it all so much.

P: Being born and raised as a Hare Krishna, did you feel any culture shock because of the difference between the world view described in the *Purāṇas* and the views of modern science?

N: No, not at all, I wasn't so versed in the details of the Fifth Canto. And there were no considerations for me as to whether or not I trusted the science. But my college courses revolved strictly around the concepts of aviation and not so much those of evolution and things like that.

P: When did you start actually flying?

N: I started flying in my freshman year and did my first solo flight after 15 hours of training. It was so exhilarating that I was singing and howling! Then eventually I became a flight instructor myself, and I was the one sending out my own students for solo flights!

P: While you were in the aviation school?

N: Yes. I became a flight instructor in my senior year before I even graduated, and I did that for two and a half years.

P: What was your career path to becoming an airline pilot?

N: That took all together 10 years. After being a flight instructor, I got hired at a cargo company transporting UPS freight at night in both the Cessna Caravan C-208 and the Embraer 110.

The E-110 happens to be a twin engine turboprop normally configured for 19 passengers, but in this case the seats were replaced by dividers and cargo straps. Because it was configured for cargo it had an exemption that allowed it to be flown by a single pilot instead of a crew of two. Little did I know this would expedite my career path by many years because Pilot In Command time in a twin engine turbine aircraft is the most qualifying flight time you could hope for to get hired at the airlines. I flew that airplane for three years and racked up 1000 hours in it, enough that I was hired directly into the captain seat at a regional Airline where I flew for Delta Connection and United Express. I was hired onto the De Havilland Dash 8 based in New York Kennedy and Denver international airports where I only worked for one year, skipping the traditional 7 to 9 years at a regional airline that most pilots go through while working their way up from First Officers to Captains. After much determination and strategy in attending job fairs preparing myself for interviews at the major airlines I was hired by AirTran Airways which 5 years later was bought by Southwest Airlines. Today I have 13 years of service at the combined airline.

P: Were you living in this area all along?

N: No. When I got hired with AirNow, the cargo company, I was based in Savannah, Georgia. I moved from the college house where I lived in New York to Savannah, and I lived there altogether for six years, but was doing a lot of commuting once I joined the airlines. Airline crew members can fly on most other airlines for free as long as there is space available. And because you'd better be flexible and willing to take whichever job offer comes to you, 50% of airline crew members don't just drive to work – they fly to work! While I lived in Savannah I commuted in this rather stressful and tiring way to my New York, Denver, and Atlanta bases. We moved to Alachua in 2011 looking to be closer to our devotee friends and the temple community and eventually, in 2012, the dream came true. I had enough seniority to hold the Orlando base with Southwest, and for the first time in a long time I was driving to work again.

P: How did you first encounter the flat earth theories?

N: It was actually some years ago when my dad came to visit, and one day we went to spend some time with a close devotee friend of his who brought up the topic and shared some of the arguments going around about the flat earth. From his point of view this was in line with the Vedic perspective, and he asked me questions on what I knew as a pilot. Actually, a lot of devotees in the community were asking me similar questions about the flat earth at the time, I guess there must have been an upsurge of YouTube videos on the topic. I became curious myself, and started looking at the information out there as well. There were certainly some compelling arguments that I didn't have answers to or didn't know how to interpret as well as some others that clearly weren't well-founded, based on what I knew as a pilot. In general, I don't like to take positions on controversial theories like this one or the moon landing and so on. I am just curious and open to what's being discussed out there. But after watching some of the videos, I have to admit that some of the arguments had me contemplating certain concepts I had learned about in my early aviation courses that had to do with the earth being a globe. Things like compass deviation errors, weather patterns, and certain aviation routes.

S. E. Kreitzer: How do you relate all this to what you're seeing as a pilot?

N: Well, I currently fly the Boeing 737, which at my airline can hold up to 175 passengers, a crew of 4 flight attendants, and 2 pilots. The cockpit has six windows, about 35 inches each, giving its pilots about a 210° field of view, and it is certified to fly up to 41,000 feet – that's 12,500 meters. At 36,000 feet and on a clear day, which was the case on my last flight, I felt like I could see a slight curve of the horizon, and it was slightly more pronounced once we reached 41,000 feet.

Several years ago I flew with a captain who was a former U-2 pilot in the Air Force. The U-2 was created for surveillance purposes working at an altitude so high up that ground equipment from rival nations couldn't detect it. He shared with me what it was like to be up at 70,000 feet, above the upper limits of class A airspace, the highest level in the national airspace system that extends from 18,000 ft to 60,000 ft, literally on the verge of outer space. He shared how the sky was black above him, and the clear definition of the earth's curvature that he observed looking down from the U-2. So, by talking to other pilots and from my own experience I'm fairly convinced that looking at the

Figure 43: Emmanuel Namamrta Greene in the cockpit of a Boeing 737-800 NG at 36,000 feet on June 27, 2021. The 210º view offered by the cockpit windows suggests curvature along the horizon.

horizon from a cockpit we *do* see the curvature of the earth. I don't think it's an optical illusion.

Another interesting point to make from an aviation standpoint is that international flights will fly what are called polar routes. These routes are created to connect a departure and destination airport using the shortest distance, which in some cases is leaning towards the North Pole or flying directly over it rather than flying along the circumference of the earth as it's presented on a flat map, which would be a lot farther.

S: What are some of the questions you get from flat earth advocates?

N: One question I have been asked a couple of times is: "When you fly your airplane, do you have to pitch down to maintain altitude?" It's a good question, but in a way it's not quite relevant.

S: How so?

N: Well, there are 4 different kinds of altitudes referenced in aviation. The first is the altitude above the ground, or AGL for "above ground level." For example, there is a regulation to maintain at least 500 feet AGL above populated areas. That's measured in feet above the ground. Also, when we request the local weather at an airport, the altitude of cloud coverage (or more precisely the base of the clouds) is given to us in feet AGL because when we fly an instrument approach, we want to know what height above the ground we are going to be when we pop out of those clouds and see the runway.

The next altitude is the height above sea level calculated in feet MSL or "main sea level. This happens to be the altitude we reference on our altimeters, also known as Indicated Altitude. Below 18,000 ft. the aircraft altimeters are set to the local barometric pressure in order to be accurate in representing our altitude above sea level. So when checking in with an air traffic controller during a climb out of New York La Guardia, we may say, "New York departure Southwest Flight 6489 is with you 8000 feet climbing to 10,000." Well, that's 8000 ft. above sea level in New York is indeed close to 8000 ft. above the ground, since the airport elevation is only 21 ft. MSL. However, if I were to check in with ATC in the same scenario climbing out of the Denver airport, 8000 ft. MSL would only be 3000 ft. above the ground since the Denver airport elevation is a mile high or 5434 ft. above sea level to be exact.

Now as we fly along, maintaining 10,000 ft. for example, and enter a new control sector, the controller will give us a new local altimeter setting current to our location. Changing the altimeter setting from say 29.92 to 30.00 HG, or inches of mercury, will increase the indicated altitude by 80 ft. So in that sense we will have to descend 80 ft. when being given this new local pressure in order to maintain an accurate 10,000 ft MSL altitude assignment.

Once we climb above 18,000 ft., however, we enter into Class A airspace where terrain is no longer a concern, and all airplanes are required to set their altimeters to the standard atmospheric pressure reference of 29.92 inches of mercury (HG). This eliminates the necessity for aircrafts to constantly update their altimeter settings for the local barometric pressure, and in this condition, the altitude referenced is called pressure altitude, which no longer represents the actual height above the ground or the sea.

Then there is density altitude, which is pressure altitude corrected for nonstandard temperatures. Density altitude is used in computing the performance of an aircraft and its engines to make sure you have enough power to get off the ground before the end of the runway and to clear obstacles on the departure. A fully loaded aircraft taking off from a 5,000-foot runway in the winter may not necessarily have the capacity to do so in the summertime, because when the density altitude is high there are fewer air molecules for the engines to grab onto.

Ok, so that's a lot of information, but the point is that the way the pitch of an aircraft is determined when it comes to maintaining altitude is not technically in reference to its height above the ground. If it were, you would be on quite the roller coaster ride when passing over mountain ranges or even a small hill. I believe the assumption is that if, as pilots, we don't have to pitch down to maintain altitude as we travel around a supposedly round earth then it must obviously be flat! It comes from a good place but it's rather naive. It's just not that simple, is it?

S: I have a question: How do you handle some of these discussions when confronted by someone who wants to explore details of contemporary flat earth arguments? My understanding is that you tactfully sidestep.

N: Well, like I mentioned earlier, I'm not really one to try to prove or disprove scientific theories or the many controversies out there. I have other passions in life. But I do enjoy learning about them, and I am usually pretty open to possibilities. It's part of my love of flying; I like discovering new concepts, different laws of nature, and how things work.

S: That's natural. Here we are human beings looking at the world around us, attempting to make sense of what it is all about. That's natural for any thoughtful person. Speaking of which, now I'd like to ask you to also talk a little bit about your relationship with your dad and your uncle.

N: Well, my dad, Yogesvara or Joshua Greene, and his brother Brian Greene, are both very well respected in their individual fields. And they have an interesting relationship because of their family history, their journey together as siblings, and what might seem like conflicting careers and views on life and creation. When my father joined the Krishna temple in London, he was 19, and it naturally affected his three younger half-siblings, including Brian, who remained in New York while his older brother spent 13 years of monastic life in Europe and other parts of the world.

They have maintained a respectful, loving relationship and have gotten much closer in recent years.

S: Concerning your uncle, there's been a lot written on Brian Greene.

N: Yes, my uncle Brian is actually a world renowned physicist. He specializes in string theory and the study of the cosmos through quantum physics.

S: He is an internationally known celebrity scientist, if you will.[1] Do you have an active personal relationship with him as his nephew?

N: I see uncle Brian every now and then at family gatherings: Thanksgiving, he was there for my wedding reception, and things like that. I honestly wish I had more opportunities to spend time with him, but we live in different states and I haven't been able to see him as much as I would like. But over the years, and hearing from my dad about his relationship with him, I have been quite fascinated by what unites these two brothers, to the point that I sincerely believe a publicized discussion between the two of them would make quite the event for those interested in cross-disciplinary fields of research.

After my dad entered family life, he eventually got into filmmaking, making different kinds of documentary films and eventually entering deep into the topic of the Holocaust, all the while maintaining an unflinching conviction and dedication to the study and teachings of the *Gita* and Vedic wisdom. He was once a teacher of Eastern philosophy at Hofstra University, has a popular podcast called Gita Wisdom, and lectures weekly in some of the top yoga studios in New York City. He has also been involved with the Bhaktivedanta Institute for Higher Studies (BIHS), and was a board member for the museum portion of the Temple of the Vedic Planetarium (TOVP) in Mayapur, India, joining discussions how to best present the Vedic sciences to the public in this monumental temple being built in India, as you well know.

Of course, he is also quite familiar with Brian's work and often incorporates the science in his explanations of Vedic philosophy. What's so interesting is that while my dad, Joshua Greene, focuses his time on teaching in a practical and accessible language, dare I say, the religious or God conscious approach to life according to *Gita* philosophy and Krishna Consciousness, his brother, Brian Greene, is world famous for presenting his research on fundamental questions of life from a strictly scientific approach. From what I observe, they are both pursuing the same big questions of life from different angles.

That's the sense that I get from both of them. Through their careers, their individual fields of research, and in their personal journeys they are deeply invested in a very similar goal while doing it from what is traditionally presented as two opposing views. They seem to personify the age-old dichotomy between science and religion, yet I find that they actually complement each other very well, and I can see this in the utmost respect my father and uncle have for one another. I've never witnessed any kind of conflict between the two of them. Actually, our Greene family members all have a close loving relationships. When I went to Thanksgiving a couple of years ago, I hadn't been with that side of my family in a really long time. But I was welcomed with so much love it almost caught me off guard, and it was really nice to see how my dad and his brother interacted in person. A very cool duo, knowing the depth of what they do. So, I feel really lucky that I have this example in my life, and to be part of a family of such intellectuals offering different pieces of the puzzle to the understanding of life and creation.

S: That's great. Your uncle has also authored a number of international best sellers, such as *The Fabric of the Cosmos* (2004) and *The Elegant Universe* (2000). Topics discussed in these works were also explored in well-received PBS *Nova* television series.

N: Yes, he's also been interviewed numerous times on television, he appeared on David Letterman, and was a guest star on the TV show, *The Big Bang Theory*. He's done all kinds of interviews, including the Joe Rogan Podcast more recently.

S: As far as scientists go, he is a very well-known public intellectual. And from what I have seen, both brothers are at ease speaking about each other in their public presentations, as you have been describing. They are mutually respectful.[2]

N: Absolutely. And from what I can gather, Brian acknowledges the possibility that what he is studying and presenting could ultimately be defining the actions of a superior intelligence. But he draws a very strong line that this will not be part of *his* process, even though that could be the conclusion. It's not part of his MO to include anything that doesn't fall in with the scientific process.

S: As a disciplinary standard, professional science upholds science-specific criteria which avoids bringing in metaphysics. But as human beings, and since all scientists are human beings, they also tend to ask those questions. So as a human being, a scientist may well be open to that, but as a practicing professional scientist, he doesn't bring that up.

N: Right. Exactly. Then you have my dad, who specializes in the Vedic texts and presents another kind of process of knowledge, which you could call the science of life: when you reach the limits of your telescope, you have no choice but to look inwards. And through the process of humility and living a life of goodness and accepting that our intelligence can't fully understand everything, that qualification itself is what can make us understand these deeper parts of creation. I see my father's approach as also scientific, but not in the traditional sense, which is based solely on matter.

So, that's how I observe my father and my uncle. They have an interesting harmony between them, and I just find the whole thing really fascinating. I personally believe both approaches are necessary to make real progress. Again, I just consider myself lucky to be close enough to witness this and learn.

S: Sounds like they share an affectionate loving relationship.

N: Yes.

S: How do you consider your own spiritual journey?

N: Well, I'm convinced of a certain statement I once heard my father say in a lecture: "The human condition is that we are eternal souls trapped in a temporary body." That conviction, along with the philosophy of Krishna Consciousness I grew up with and the high esteem I hold for many exemplary devotees in my life, have laid a pretty solid spiritual foundation for me. Enough that I do view my life according to the principles of *bhakti* and have faith in whatever parts of the tradition I may not fully understand yet. The main thing that's been on my mind for many years now is how to practically apply the standards and practices pushed by the institution in ways that are realistic yet uncompromising for the next generations of devotees. I think we have some progress to make in making ISKCON more relevant to all areas and demographics of an evolving world. And what you are doing here, is I think, a great contribution. What's often highlighted is usually very much associated with the early lifestyle of first generation devotees who were living a monastic life in temples as monks and preachers. I can't necessarily do a lot of those things, so how do I make my current life circumstances an integral part of my spiritual journey? Like everyone I often find myself in the "struggle for existence," as Srila Prabhupada calls it – whether it's the lifestyle of being an airline pilot, constantly on the road trying to maintain good health while having to eat a lot of airport food, or being in a married relationship and raising a disabled child. I know deep inside these life circumstances have been custom-tailored for me by Krishna Himself, so I'm always trying to introspect and understand: What does He want me to do? How am I meant to process this?

P: Thanks for sharing all this with us today. It's been great hearing about your career and your personal experiences as a pilot. Do you have any concluding words: some message you want to present or just some feeling that you have about this?

N: Well, first of all I want to thank you for inviting me to do this. It's really an honor, and I find it very exciting to know a little more about what you all do at the BI. I think what we're doing here is inviting people into the all-important conversation of trying to understand the creation, life on this planet, and where we fit into all of it. And what I appreciate about this talk, this interview, is that you've given my career as a pilot a significance in a bigger picture as part of this broader discussion. I want myself to become expert in inviting people into that conversation.

The US airline industry holds some interesting world records. The Atlanta Hartsfield Airport is the busiest airport in the whole world, Denver International has the largest footprint, and Southwest Airlines happens to carry the most passengers on any given day. In order to do so it employs close to 10,000 pilots. So in my career, I am in a unique position where I usually get to meet someone new every week and spend some 20 to 30 hours with them in a small cockpit. A new and unique individual to be discovered and with whom to potentially share some fun and enlivening conversations. From my numerous experiences in doing this over the years, and trying to refine my ability to engage in meaningful conversations, I find that when the walls of insecurity, ego, and social discomfort are taken down, most people are very intrigued by our philosophy and love to connect and explore on a deeper level.

APPENDIX 9

Personal Testimony: "Experiments That Can Be Done at Home"

By Prahlad Griffith

I was born into the Hare Krishna movement in 1991, but like many other devotees, my parents moved around a lot. I lived the first three years of my life in Dallas, Texas, before we moved to Denver, Colorado, where my father, Janmastami Prabhu, became the temple president. My brother Devala was born there, and we stayed until I was about five, when my parents decided to move to India. We spent six months in Vrindavan and then decided to move to Mayapur. I attended school there over the next seventeen years. I would occasionally travel to different places in India – Vrindavan, Jagannath Puri – as well as to the United States. Then in 2013, we moved back to the US where I started teaching *mrdanga* and harmonium. I now have a website, http://kirtanstudios.com, where we offer classes in the musical arts drawn from the *bhakti* tradition.

Growing up in Mayapur, I somehow always seemed to love math and science. I was quite good with that kind of thinking – mathematical, logical thinking. I completed 10th grade math, physics, and chemistry courses at home on my own in three months, and scored the highest in India on the tenth grade physics exam. That kind of showed me that I had some strength in the area. So that is how I got interested in science. I would always read books or watch documentaries on different things like that, about science.

At the same time, while living in Mayapur, I was exposed to a number of conspiracy theories since a lot of people there were getting into them during that period. For example, I seriously considered Kent Hovind's young-earth creationist arguments, and for a while I adopted his anti-mainstream agenda. I would also embrace various exotic alternative healing modalities. But when two senior devotees died from tuberculosis after unfortunately drinking contaminated raw milk and refusing standard health remedies, I began to rely more on evidence-based models. I experienced many of them as being beneficial and sometimes even necessary for eliminating suffering. That furthered my interest in science and evidence-based models.

Then when I moved to Alachua, Florida, and began teaching music, I began hearing from a lot of teachers and parents discussing flat earth conspiracies. A number of the children and a few teachers appeared to consider them seriously. I felt frustrated because it didn't make sense to me on so many levels. I found that debating the issue with logic and reason was not working. For example, I could reference *śāstra* but then they had other references that they felt supported their position, so it wasn't going anywhere.

At one point I decided I would do something that could prove this whole thing wrong, or at the least, get to the bottom of it. I decided to use a very simple experiment based on a *National Geographic* video that undermined flat earth arguments. The idea was: why do lighthouses have to be high up? It is expensive to do that. If the earth was flat, a lighthouse would only need to be a few feet tall, just above the waves, and people should be able to see its bright light hundreds of miles away. Why then, for all these years, have they been building huge towers, when if the earth were flat, they could have made them a few feet tall?

My friend Varuni Sutton and her ten-year-old son Ramai felt inclined toward the idea that the earth was flat, but they agreed to help me with the experiment. Another friend, Radha Priya, who has a master's degree in chemistry, joined us. And my brother Devala decided he would come too. So that was our crew. We all drove to Lake George at Silver Glen Springs in Ocala, Florida. The lake is about six miles wide, and has a pier. We brought some lights that we would set up on either side of the lake, along with our cameras and phones to film it.

Then we raised and lowered the lights. The idea was that if the earth were flat, then the lake would be flat, too. A lot of flat earth conspiracy

Figure 44: The author, Prahlad Griffith, who along with friends prepared for some serious science along Lake George at the Silver Glen Springs Recreational Area in Ocala, Florida.

advocates would say that you should be able to see lights clear across the lake if they were bright enough, no matter what height we placed them. So, if you took your bright light and put it pretty close to the water, you should be able to zoom in with a telescope and still be able to see it from the other side six miles away.

We had a pole, and we taped a giant flashlight to it. I don't recall exactly, but I think it was 900 lumens, a really powerful light. Now our experience was that we had to raise the light about 20 feet above water level in order to see it from across the lake. Of course, this also depended on the height of the viewer, but it was about 20 feet.

All the details are in the video we took. I have it posted on YouTube at the following address: https://youtu.be/TaxImroFj7g. In it, I show my equipment and explain how we did the experiment, how we brought the light up and down. When we brought the light up, it would become visible, and when we dropped it down, it would disappear. We just kept doing that, and we basically found the horizon by doing this simple experiment.

The people with us who had previously considered flat earth perspectives, changed their mind after coming out there and seeing the experiment. Although they were somewhat persuaded beforehand by the various counter arguments based on a spherical earth model, when we did the simple experiment with the light disappearing and reappearing at 20 feet, they were totally convinced!

Varuni, I think she was almost convinced even before during our drive to the lake, maybe because I was so confident that we would get a certain result. She brought up all sorts of reasons why the earth might be flat, and I kind of debunked them. Though she had been arguing for a flat earth, she was open to experimenting, and that made the difference for her.

Afterwards, I told a few people about this experiment, and it was like if they didn't see it, they really didn't believe it. So perhaps the best way to convince people would be videoing an experiment, making it colorful with good editing while adding humor. Things like that, having something dramatic going on that is appealing to people, entertaining and fun. Keep it visual, something simple they can relate to. A project like that might also interest young people to investigate the utility of practical science.

APPENDIX 10

ISKCON Art During the 1970s: Earthly Spheres in Prabhupāda's Books

by Prishni Sutton and S. E. Kreitzer

The following collection of ISKCON BBT artwork was commissioned during Prabhupāda's lifetime. All of these images were published in early editions of his books, with many remaining in present editions. A number of them clearly depict earthly spheres familiar to contemporary audiences. Undoubtedly, Śrīla Prabhupāda was well aware of the numerous representations of the earth as a globe utilized on behalf of his outreach efforts. As noted by a veteran BBT artist (see description for Figure 8), Śrīla Prabhupāda approved each painting "if not before, definitely after" completion prior to the publication of his books.

There is no a clear record of Prabhupāda expressing concern for presenting the earth as a spherical object. To be sure, the earth as a globe was also a perspective held within the *jyotiṣa* tradition integral to Vedic civilization, broadly speaking (see Chapter 4), as well as Purāṇic literature specifically. This is evident from references of *Jyotiḥśāstra* consulted within the *Purāṇas* to plan religious functions and gain insight into temporal affairs. After all, the *Bhāgavatam* considers the sage Gargamuni "a great authority" as a *jyotiṣa paṇḍita* (SB 10.8.6, purport), as well as the "*ācārya* of the Yadu dynasty" (*Kṛṣṇa* book, Ch. 45). He played an integral role in the Tenth Canto intimately advising Nanda Mahārāja and Yaśodā-mayī shortly after Kṛṣṇa's birth. Nanda Mahārāja directly welcomed Gargamuni as "the best of the *brāhmaṇas*, especially because you are fully aware of the *jyotiḥ-śāstra*" (SB 10.8.6).

In the *Śrī Caitanya-caritāmṛta* as well, Prabhupāda mentioned in a conversation (LA, 08/04/69) that Nimai's "grandfather was a great astrologer, His mother's father, Nīlāmbara Cakravartī. He immediately calculated His horoscope, constellation of the stars under which He was born. He calculated that 'This boy will be a great reformer.'" It is useful to remember that in traditional culture, not only in India but around the globe, mathematics, astronomy, and astrology were intrinsically related in a manner similar to Prabhupāda's remark that "in the M.A. class, the student is studying higher mathematics, astronomy, astrology, there also, the two plus two equal to four is the truth" (BG 05.03–7, NY, 08/24/66).

Also of note, Prabhupāda's spiritual master, Śrīla Bhaktisiddhānta Sarasvatī, was recognized as a leading scholar of Sanskrit mathematics and *jyotiṣa* astronomy, disciplines he actively utilized until he took *sannyāsa*. With regards to the Purānic tradition, it was Bhaktisiddhānta Sarasvatī Ṭhākura's Bengali translation with commentary on the Fifth Canto of the *Śrīmad-Bhāgavatam* that Śrīla Prabhupāda drew upon heavily while writing the English language BBT edition of the Fifth Canto. Suffice it to say, the great Vaishnava *ācāryas* knew both traditions.

Each of the images presented in this section include publication dates along with information on the artists as most recently acknowledged by the Bhaktivedanta Book Trust (BBT).

Figure 1: *Easy Journey to Other Planets*
(Boston: ISKCON Press, 1970)
Cover Illustration: artwork by Jaduraṇī dāsī

Śrīla Prabhupāda's first book, *Easy Journey to Other Planets* (1960), was drawn from two essays that he initially published in *Back to Godhead* magazine in 1959. A positive letter written in response by the Indian physicist, Dr. Y. G. Naik inspired Bhaktivedanta Swami to publish these essays in a book that included an account of their written exchanges.

This image on the cover of the first American edition was published in 1970 by ISKCON Press. Amongst the many celestial globes, the planet earth, with visible continents, appears in the "material realm" on the left, with the devotee's journey toward the "spiritual realm" on the right. The same image is included as Plate 3 in the 1972 BBT edition of the work.

A description titled "On the Cover," is located toward the end of the book just before the glossary. It states:

> The picture shows the comparative progress of a material spacecraft on its way to the moon and other nearby planets and a pure spirit soul journeying to the spiritual planet Goloka Vṛndāvana, located far, far beyond these material universes. The painting is not imaginative, but is based on scientific fact and authentic scriptural reference.
>
> Certainly the velocity of material rockets has not even approached the speed of light, whereas the spirit soul can travel much faster than the speed of the mind. It is said in the *Brahma-saṁhitā*: "Just ride for millions of years on the airplane which runs at the speed of mind, and you will find that the spiritual sky is unlimited." Not only do they lack the speed to approach Goloka, but the material scientists have no information of the destination where life is eternal, blissful and full of knowledge.
>
> According to the *Bhagavad-gītā* [8.16], "From the highest planet in the material world, down to the lowest, all are places of misery, where repeated birth and death take place." There are particular descriptions in the Vedic literature of the vast, eternal spiritual sky filled with innumerable planets where the presiding deity is the Supreme Personality of Godhead and where life is free of all anxiety, full of bliss and knowledge.
>
> Only by pure unalloyed devotional service to God, or Kṛṣṇa, can the spiritual soul become free from its material covering and in less than a second transfer to its original spiritual form (*svarūpa*) in the spiritual sky. The most subtle and powerful force in existence is called *bhakti-yoga*, or love of God, and the principle by which anyone can make the

journey back to home, back to Godhead, is the practice of transcendental loving service to Kṛṣṇa, as taught in the pages of *Easy Journey to Other Planets.*

Figure 2: *Kṛṣṇa: The Supreme Personality of Godhead* (Vol. 1)
(Boston: ISKCON Press, 1970)
Artwork by Baradrāja dāsa

That same year, ISKCON Press published the first edition of *Kṛṣṇa: The Supreme Personality of Godhead*, financed by a generous donation from George Harrison of The Beatles. All the celestial objects in this painting appear as globes in space.

The published caption for the image reads: "Mother Yaśodā saw within the mouth of Kṛṣṇa the whole sky, including all the stars and all the directions. (p. 54)"

The following recollection from the *Srila Prabhupada–Remembrances* series by Siddhānta dāsa, specifically refers to this painting:

> **Jahnava:** Śrīla Prabhupāda called Baradrāja and me to his room to talk about painting for the *Kṛṣṇa* book. Baradrāja

was already speaking to Śrīla Prabhupāda when I came in. Śrīla Prabhupāda began telling *Kṛṣṇa* book stories we had never heard before ...

Then Śrīla Prabhupāda began telling us about the Universal Form, the universe contained within the mouth of Lord Kṛṣṇa when Mother Yaśodā asked Him to open His mouth. Śrīla Prabhupāda described how, "There were oceans and seas, and there was air and fire and earth and water and the ten directions, north, south, east and west, mountains and ..." for about five minutes.

As he was speaking in this wonderfully charismatic way, Baradrāja and I, unbeknownst to us, were gradually leaning forward, getting involved in the story. Then at one point, Śrīla Prabhupāda said to Baradrāja, "Now you paint that!"

It was such a shock. It was a double shock because we were coming closer and closer and we weren't expecting Śrīla Prabhupāda to suddenly say that, and it was also a shock to be asked to paint it. Baradrāja almost lost his balance, and I fell backwards too.

Months later, Baradrāja was in Montreal struggling with how to paint this picture. He wrote to Śrīla Prabhupāda and said, "Śrīla Prabhupāda, in all due honesty, I cannot figure out how to put all those things in Kṛṣṇa's mouth." Śrīla Prabhupāda sent back a little sketch that showed a stick figure of Mother Yaśodā standing up, a little stick figure of child Kṛṣṇa sitting down with a balloon coming out of His mouth, and Prabhupāda said, "That's how you do it." Then Baradrāja made the beautiful painting that's in the original *Kṛṣṇa* book of the universe in Kṛṣṇa's mouth. That painting now hangs in the foyer of the Montreal temple.

(For more information, see:
www.srilaprabhupadalila.org/read/SPFPP_0290)

Figure 3: *Śrīmad-Bhāgavatam,* Second Canto
"The Cosmic Manifestation" Part One, Chapters 1–6
(Los Angeles: Bhaktivedanta Book Trust, 1972)
Plate Five: artwork by Jaduraṇī dāsī

The original 1972 BBT edition of the *Śrīmad-Bhāgavatam,* Canto 2, Part 1, includes Nārada Muni traveling through space surrounded by spherical planets. Of note, it is generally acknowledged that the first account of the rings of Saturn occurred in 1610.

The published caption for Plate Five references SB 2.2.23 and states:

> Because he is fixed in unbroken devotional service to the Lord, Nārada Muni has received the grace of the almighty Viṣṇu. Thus he is a liberated spaceman, and he can travel anywhere without restriction, both in the transcendental world and in the three divisions of the material world. As he travels, he constantly sings the glories of the Lord and vibrates his *vīṇā*, which is charged with transcendental sound and which was given to him by Lord Kṛṣṇa Himself. (*p. 106*)

Figure 4: *Śrīmad-Bhāgavatam,* Second Canto
"The Cosmic Manifestation" Part Two, Chapters 7–10
(Los Angeles: Bhaktivedanta Book Trust, 1972)
Plate One: artwork by Baradrāja dāsa, Murlidhara dāsa, Jadurāṇī dāsī

In Plate One of the second volume of the original Second Canto published by the BBT, a globular earth can be seen floating on the Garbhodaka Ocean near the Hiraṇyākṣa's legs.

The published caption for Plate One references SB 2.7.1 and states:

> The Lord assumed the gigantic form of a boar and fought with the demon Hiraṇyākṣa. (*p. 338*)

Figure 5: *Śrīmad-Bhāgavatam,* Second Canto
"The Cosmic Manifestation" Part Two, Chapters 7–10
(Los Angeles: Bhaktivedanta Book Trust, 1972)
Plate Six: artwork by Viṣṇu dāsa and others

In Plate Six of the second volume of the original Second Canto published by the BBT, the *avatāra* Lord Vāmanadeva, who "covered the entire surface of the earth with one footstep" (SB 8.20.32–33), is shown with one foot on the surface of the earth globe as his first step, his second step stretching to the limits of the universe, and his third step touching the head Bali Mahārāja.

The published caption for Plate Six references SB 2.7.18 and states:

Bali Mahārāja was very glad to receive the Lord's remaining step on his head. (*p. 366*)

Figure 6: *Śrī Caitanya-caritāmṛta, Ādi-līlā*
Vol. 2: "Lord Caitanya Mahāprabhu in the Renounced Order of Life"
(Los Angeles: Bhaktivedanta Book Trust, 1973)
Plate Eleven: artwork by Jadurāṇī dāsī, Viṣṇu dāsa, and others

Plate Eleven from *Ādi-līlā*, Volume Two of the original *Śrī Caitanya-caritāmṛta* published by the Bhaktivedanta Book Trust, illustrates the *varāha-āveśa* pastime of Lord Caitanya described in *Ādi-līlā* 10.49, that symbolized the earth as a round waterpot.

Śrīla Prabhupāda's purport for *Ādi-līlā* 17.19 further describes:

> One day Caitanya Mahāprabhu began to cry out, "Śūkara!

> Śūkara!" Thus crying out for the boar incarnation of the Lord, He assumed His form as the boar incarnation and got up on the shoulders of Murāri Gupta. He carried a small *gāḍu,* a small waterpot with a nozzle, and thus He symbolically picked up the earth from the depths of the ocean, for this is the pastime of Lord Varāha.

The published caption for Plate Eleven references CC *Ādi-līlā* 10.49 and states:

> Lord Caitanya exhibited His Varāha form in the house of Murāri Gupta, the twenty-first branch of the Śrī Caitanya tree. (*p. 279*)

Figure 7: *Śrīmad-Bhāgavatam*, Third Canto
"The Status Quo" Part Three, Chapters 17–21
(Los Angeles: Bhaktivedanta Book Trust, 1974)
Plate One: artwork by Parīkṣit dāsa and others

As with the image in Figure 4 published two years earlier, Figure 7 likewise shows an earth-like ball floating on the ocean, along with another situated on Varāha's shoulder.

The published caption for Plate Six references SB 3.18.18 and states:

> The demon and the Lord struck each other with their huge maces, each enraged and seeking his own victory. (*p. 715*)

Figure 8: *Śrīmad-Bhāgavatam,* Fifth Canto
"The Creative Impetus" Part Two, Chapters 14–26
(Los Angeles: Bhaktivedanta Book Trust, 1975)
Plate Three: artwork by Gaurī dāsī, Haridāsa
(BTG 11.11 (1976) credits Parīkṣit on p. 26)

This painting has particular significance as it was used to illustrate verses in the cosmology section in the Fifth Canto of the *Śrīmad-Bhāgavatam.* There can be little doubt that the sphere resting on the tusks of Lord Varāha represents the earth, since the continents of the Western Hemisphere are clearly visible. In addition, at the top center just above the wing of one of the celestial swan airplanes, is an image suggestive of a galaxy.

Puṣkara dāsa (Matthew Goldman), the long-time ISKCON BBT artist, noted in an oral history interview with Prishni Sutton and S. D. Muni, held on October 31, 2021 in Alachua, Florida, that Śrīla Prabhupāda approved each painting "if not before, definitely after" completion, prior to publication in his books. Puṣkara added, "The idea that Prabhupāda didn't see it [a painting], or didn't approve of it is out of the question. He saw."

During a Vṛndāvana room conversation held on June 28, 1977, Prabhupāda discussed the portrayal of Fifth Canto constructs with his disciples (Conv. D, 7770628R4). Puṣkara inquired:

> **Puṣkara:** Śrīla Prabhupāda, previously we painted in the art department ... It's described that Varāha lifted the earth, and the earth was a globe, and we showed also a globe of the earth. How does that relate to this? Previously, when we painted, we showed the earth a ball. So now the artists will be very confused. How it fell in the Garbha Ocean as a ball?
>
> **Yaśodānandana:** It depends on what we mean by earth. The Western conception of earth is just five continents and a few oceans, but according to *Bhāgavatam*, earth means Jambūdvīpa, because earth is connected with Jambūdvīpa.
>
> **Puṣkara:** So whole Jambūdvīpa fell ...
>
> **Prabhupāda:** [No,] I have already [explained,] ... earth globe.

Prabhupāda would add toward the end of the discussion:

> **Tamāla Kṛṣṇa:** That is what we're painting next ... Bhū-maṇḍala, the whole Bhū-maṇḍala, these different oceans and *dvīpas*. That will be in color.
>
> **Prabhupāda:** Very nice.
>
> **Tamāla Kṛṣṇa:** This Puṣkara, he did the ...

Puṣkara: I made the mess. (laughs)

Prabhupāda: He'll do something wonderful in arts. Kṛṣṇa will make. They will combine. All right. *Jaya.*

In October 2021, almost a half-century later, Puṣkara would offer the following recollection of his personal exchanges with Śrīla Prabhupāda:

Puṣkara: So, that time, somehow I was just there with Prabhupāda. I just said, "Prabhupāda, nobody can see Sumeru or any of the things that we're making maps of. We would see it if it's so big." I said, "It must be subtle."

And he said, "Yes, it's subtle."

Then I thought, okay, let me ask him something else. And I said, "Well, but the way we're reading about Bhū-maṇḍala, it doesn't exactly sound like a globe, but we always show a globe with the Boar and it says, *gola*," and I threw in the Sanskrit word.

And then Prabhupāda said ... He didn't say anything. He just looked at me like this ... Just kept doing whatever he was doing. He's totally like ...

S. D. Muni: Like he knew what you were saying, but didn't want to respond?

Puṣkara: Like I answered my own question. Like: "You idiot, yeah, we show a globe." That's the way I got it.

Apparently, Prabhupāda would express little to no concern portraying the earth as a globe. After all, this was a perspective accepted within the respected *jyotiṣa* tradition integral to Vedic culture, as well as in the Western traditions of natural philosophy and professional science. At the same time, Śrīla Prabhupāda actively promoted the authenticity of the cosmographical accounts in the Fifth Canto of the *Śrīmad-Bhāgavatam*. He appeared to feel that the sincere intentions of his followers combined with the desire of Lord Śrī Kṛṣṇa would accomplish something wonderful, while exploring the complexity of *Virodha-parihāra* discourse.

The published caption for Plate Three references SB 5.18.35–39 and states:

> The Supreme Lord in His boar incarnation, who accepts all sacrificial offerings, lives in the northern part of Jambūdvipa. There, in the tract of land known as Uttarakuru-varṣa, mother earth and all the other inhabitants worship Him with unfailing devotional service by repeatedly chanting the following Upaniṣad *mantra*: O Lord, we offer our respectful obeisances unto You as the gigantic person. Simply by chanting *mantras*, we shall be able to understand You fully. You are *yajña* (sacrifice), and You are the *kratu* (ritual). Therefore all the ritualistic ceremonies of sacrifice are part of Your transcendental body, and You are the only enjoyer of all sacrifices. My Lord, as the original boar within this universe, You fought and killed the great demon Hiraṇyākṣa. Then You lifted me (the earth) from the Garbhodaka Ocean on the end of Your tusk, exactly as a sporting elephant plucks a lotus flower from the water. I bow down before You. (*pp. 212–220*)

Figure 9: *Śrī Caitanya-caritāmṛta, Madhya-līlā*
Vol. 8: "The Lord's Teachings to Śrīla Sanātana Gosvāmi"
(Los Angeles: Bhaktivedanta Book Trust, 1975)
Plate Six: artwork by Baradrāja dāsa, Jadurāṇī dāsī,
Muralīdhāra dāsa

The 1975 painting in Figure 9 not only depicts the universal globes in the lower half of the painting emanating from the gigantic body of Mahā-Viṣṇu, but also just above Mahā-Viṣṇu, the individual planets as globes situated within each of the universal globes.

The published caption for Plate Six references CC *Madhya-līlā* 20.278–89 and states:

The first form of Lord Viṣṇu is called Mahā-Viṣṇu. He is the original creator of the total material energy. The innumerable universes emanate from the pores of His body. These universes are understood to be floating in air as the Mahā-Viṣṇu exhales. They are like atomic particles that float in sunshine and pass through the holes of a screen. All these universes are thus created by the exhalation of Mahā-Viṣṇu, and when Mahā-Viṣṇu inhales, they return to His body. The unlimited opulences of Mahā-Viṣṇu are completely beyond material conception.

After creating the total number of universes, which are unlimited, the Mahā-Viṣṇu expanded Himself into unlimited forms and entered into each of them. When Mahā-Viṣṇu entered each of the limitless universes, He saw that there was darkness all around and that there was no place to stay. He therefore began to consider the situation. With the perspiration produced from His own body, the Lord filled half the universe with water. He then lay down on that water on the bed of Lord Śeṣa. A lotus flower then sprouted from the lotus navel of that Garbhodakaśāyī Viṣṇu. That lotus flower became Lord Brahmā's birthplace. In the stem of that lotus flower the fourteen worlds were generated. Then He became Lord Brahmā and created the entire universe. In this way, the Supreme Personality of Godhead in His form of Viṣṇu maintains the entire material world. Since He is always beyond the material qualities, the material nature cannot touch Him. (*pp. 157–162*)

Figure 10: *Śrīmad-Bhāgavatam,* Eighth Canto
"Withdrawal of the Cosmic Creation" Part Three, Chapters 17–24
(Los Angeles: Bhaktivedanta Book Trust, 1976)
Plate Four: artwork by Jayarāma dāsa

Four years after the publication of a painting of Vāmanadeva taking three steps of land (Figure 5), a second image portrayed Him with feet resting on a spherical planet (Figure 9). Prabhupāda's translation of SB 8.20.23 mentions that Bali Mahārāja "saw on the Lord's feet the surface of the globe."

The published caption for Plate Four references SB 8.20.21–33 and states:

As Bali Mahārāja and the members of the assembly watched in stunned amazement, the unlimited Supreme Personality of Godhead, who had assumed the form of Vāmana, began increasing in size, until everything in the universe was within his body, including the earth, the planetary systems, the sky, the oceans, the birds, beasts, human beings, the demigods and the great saintly persons. The Lord in His universal form thus stood before Bali Mahārāja holding a conchshell, sword, shield, flaming discus, arrow, bow, lotus flower and club. On the Lord's feet, Bali Mahārāja saw the surface of the globe, on the surface of His calves all the mountains, on His bosom all the clusters of stars as well as the goddess of fortune, who held a lotus flower in her hand, and on His hair, the clouds. The Lord wore a yellow garment, covered by a belt, and He was decorated by a flower garland, surrounded by bees. Manifesting Himself in this way, the Supreme Personality of Godhead, whose activities are wonderful, covered the entire surface of the earth with one footstep, the sky with His body, and all directions with His arms. (*pp. 116–126*)

Figure 11: *Śrīmad-Bhāgavatam,* Tenth Canto
"Liberation" Part One, Chapters 1–8
(Los Angeles: Bhaktivedanta Book Trust, 1977)
Plate Three: artwork by Sulakṣmaṇā dāsī

Figure 11 illustrates the mystic sage Durvāsā Muni fleeing from Lord Kṛṣṇa's personal weapon, the Sudarśana *cakra.*

An intriguing feature found in Figure 8 and repeated again here in Figure 11, as well as the next image, Figure 12, is the presence of a spiraling celestial object resembling a galaxy. As with earlier paintings such as the 1970 cover for *Easy Journey to Other Planets,* this suggests that during Prabhupāda's lifetime ISKCON BBT artists felt comfortable portraying not only a spherical earth planet, but a variety of features identified with modern cosmology.

The published caption for Plate Three references SB 9.4.49–52 and states:

> Upon seeing that his attempt to curse King Ambarīṣa had failed and that the Lord's Sudarśana *cakra* was moving toward him, Durvāsā Muni became very frightened and began to run in all directions to save his life. As the blazing flames of a forest fire pursue a snake, the disc of the Supreme Personality of Godhead began following Durvāsā Muni. Durvāsā Muni saw that the disc was almost touching his back, and thus he ran very swiftly, desiring to enter a cave of Sumeru Mountain. Just to protect himself, Durvāsā Muni fled everywhere, in all directions – in the sky, on the surface of the earth, in caves, in the ocean, on different planets of the rulers of the three worlds, and even on the heavenly planets – but where he went he immediately saw following him the unbearable fires of the Sudarśana *cakra*. (*pp. 126–128*)

Figure 12: *Śrīmad-Bhāgavatam,* Tenth Canto
"The Summum Bonum" Part Two, Chapters 6–12
(Los Angeles: Bhaktivedanta Book Trust, 1977)
Plate Seven: artwork by Jadurāṇī dāsī, Sulakṣmaṇā dāsī,
Dhṛti dāsī, Rāmadāsa Abhirāma dāsa

Figure 12 is an updated version of the 1970 illustration (Figure 1) of Mother Yaśodā seeing the universe within Kṛṣṇa's mouth. Along with the complete cosmic manifestation that includes a number of planetary spheres, Yaśodā saw "herself and Vṛndāvana-dhāma" portrayed in this painting as a globe above Kṛṣṇa's head. As noted, Figure 12 also features a spiraling image resembling a galaxy.

The published caption for Plate Three references SB 10.8.32–39 and states:

> One day, while Kṛṣṇa was playing with His brother, Balarāma, and His other small playmates, all [of] Kṛṣṇa's friends came together and complained to mother Yaśodā. "Mother," they submitted, "Kṛṣṇa has eaten some earth." Upon hearing this from Kṛṣṇa's playmates, mother Yaśodā picked Kṛṣṇa up and said, "Dear Kṛṣṇa, why are You so restless that You have eaten dirt in a solitary place? How is this?" Kṛṣṇa replied, "My dear mother, I have never eaten dirt. All my friends complaining against Me are liars. If you think they are being truthful, you can directly look into My mouth and examine it." When Kṛṣṇa opened His mouth wide, mother Yaśodā saw within His mouth all moving and nonmoving entities, outer space and all directions, along with mountains, islands, oceans, the surface of the earth, the blowing wind, fire, the moon and the stars. She saw the planetary systems, water, light, air, sky, and creation by transformation of false ego. She also saw the senses, the mind sense perception, and the three qualities goodness, passion and ignorance. She saw the time allotted for the living entities, she saw natural instinct and the reactions to *karma*, and she saw the varieties of bodies produced by different desires. Seeing all these aspects of the cosmic manifestation, along with herself and Vṛndāvana-dhāma, she became doubtful and fearful of her son's nature. (*pp. 117–23*)

Notes

Foreword

1 For an explanation and preliminary assessment of this hermeneutics project, see the article "The Application of Hermeneutics in ISKCON" by Urmilā Devī Dāsī, in *ISKCON Communications Journal* 13 (2022), 1–45.

Introduction

1 For more information, please see Appendix 2, "Bhū-maṇḍala, or Middle Earth," by Richard L. Thompson.

2 *jyotiṣa brāhmaṇa paṇḍita*, and various permutations of these terms, is utilized in this essay to refer to traditional practioners of Sanskrit mathematics, astronomy, and astrology, and who lived within a Vedic cultural context identified with the Indian subcontinent. Casually defined, *jyotiṣa* refers to knowledge of astronomy within a Vedic context, *brāhmaṇa* refers to a member of the priestly, intellectual class, and *paṇḍita* refers to a learned scholar.

3 According to Richard L. Thompson, the "apparent conflict in understanding the Fifth Canto is not something new. And it is not a conflict between Vedic literature and modern scientific understanding. Actually, there is a whole body of Vedic literature which describes the earth as a small globe. And this literature is what Śrīla Prabhupāda in his purports refers to as *Jyotir Veda* or *Jyotiḥśāstra*." See Richard L. Thompson, "Vedic Cosmology Seminar 1," ISKCON Philadelphia, July 1, 1986, https://richardlthompson.com/vedic-cosmology-seminar-1. See also page 2 of Thompson's *Vedic Cosmography and Astronomy* (Los Angeles: BBT, 1989).

4 For an authoritative scholarly analysis of the complexity inherent to the relationship between science and religion within a broad historical context, see: John Hedley Brooke, *Science and Religion: Some Historical Perspectives* (New York: Cambridge University Press, 1991).

5 This would also involve the perception of relative motion.

6 Christopher Z. Minkowski, "The Pandit as Public Intellectual: The Controversy over *virodha* or Inconsistency in the Astronomical Sciences," in *The Pandit: Traditional Scholarship in India*, ed. Axel Michaels (New Delhi: Manohar Publishers, 2001), 83.

7 Christopher Minkowski, "Competing Cosmologies in Early Modern Indian Astronomy," in *Studies in the History of the Exact Sciences in Honour of David Pingree*, ed. Charles Burnett et al. (Leiden: Brill, 2004), 349.

8 Kim Plofker, *Mathematics in India* (Princeton: Princeton University Press, 2009), 321.

9 Minkowski, "The Pandit as Public Intellectual," 83.

10 Christopher Minkowski, "Competing Cosmologies in Early Modern Indian Astronomy," in *Studies in the History of the Exact Sciences in Honour of David Pingree*, ed. Charles Burnett et al. (Leiden: Brill, 2004), 351–352.

11 Of note, variations of the term also appear in other areas of the broad tradition. For example, Ramanujacharya (c. 1017 CE – 1137 CE) is identified with the treatise, *Virodhi Pariharangal* (Removal of Obstacles). For an English translation of this text, see: https://mannaram.tripod.com/ramanuja/pdfs/virodhpe.pdf, accessed January 28, 2024.

12 Pushpa Kumari Jain, *The Sūryaprakāśa of Sūryadāsa*, vol. I (Vadodara: Oriental Institute, 2001), 1.

13 Minkowski, "On Sūryadāsa and the Invention of Bidirectional Poetry (*vilomakāvya*)," *Journal of the American Oriental Society* 124, no. 2 (2004), p. 329. Minkowski describes that the "*Siddhāntasaṃhitā-sārasamuccaya* is a 'compendium of essential points about the astronomical Siddhāntas and Saṃhitās,'" and adds that in the last chapter of "the text, Sūrya refers to the year current at the time of writing as Śaka 1505, or AD 1583." In an earlier essay, on page 82 of "The Pandit as Public Intellectual," Minkowski appears to identify another date for the text as "1530s".

14 Minkowski, "The Pandit as Public Intellectual," 82.

15 Virendr Nath Sharma, "Sawai Jai Singh's Hindu Astronomers," *Indian Journal of History of Science* 28, no. 2 (1993), 138.

16 Clemency Montelle and Kim Plofker, "Integrating Logarithm Concepts into Syntheses of Sanskrit, Muslim and European Astronomy in 18th-Century Jaipur," in *Science and Society in the Sanskrit World*, edited by

Christopher T. Fleming, Toke Lindegaard Knudsen, Anuj Mistra, and Vishal Sharma (Leiden: Brill, 2023), p. 175.

17 Sharma, "Sawai Jai Singh's Hindu Astronomers," 138–39. Sharma notes that "Jai Singh brought him to Amber in 1725 AD" which at the time remained the capital, suggesting that was the year Kevalarāma entered service at the royal court.

18 David Pingree, "The Purāṇas and Jyotihśāstra: Astronomy," *Journal of the American Oriental Society* 110, no. 2 (April–June 1990), 280. Also see: Minkowski, "Competing Cosmologies," 378–79.

If Kevalarāma entered into the service of Jai Singh in 1725, and with Singh passing away in 1743, an educated guess could place the text c. 1730s.

19 Minkowksi, "The Pandit as Public Intellectual," 83, fn 10.

20 Minkowski, "Competing Cosmologies," 379.

21 With reference to this parable, Hans Henrich Hock, Professor Emeritus of Linguistics and Sanskrit at the University of Illinois, translates Śaṅkarācarya's *bhāṣya* on *Chāndogya-Upaniṣad* 5.18.1: "*etaddhastidarśana iva jātyandhāḥ* – That is like people blind by birth in/when viewing an elephant." See: Hans Henrich Hock, "Philology and the Historical Interpretation of the Vedic Texts," in *The Indo-Aryan Controversy: Evidence and Inference in Indian History*, edited by Edwin F. Bryant and Laurie L. Patton (New York: Routledge, 2005), 282.

From a different perspective, there is the statement attributed to Ṛg Veda 1.164.46, dated by scholars to the second millennium BCE: "There is one reality, the wise speak of it in various ways."

Perhaps a philosophical analysis could also be explored with Plato's Allegory of the Cave.

22 For example, in his commentary to *Śrī Caitanya-caritāmṛta, Adi-līl* 3.8.

23 For more discussion in *Śrīmad-Bhāgavatam* on the purpose of describing the *virāṭ-rūpa*, please see the translations and purports to 1.3.3, 1.3.30, 2.1.22–25, 2.2.14, 3.6.4–7, 5.23.4, 5.23.9, 5.26.38–40, and 6.4.32.

24 *Śrīmad-Bhāgavatam* 5.16.3. Of special attention are also verses 5.26.38–40, located at the end of the concluding chapter to the Fifth Canto. There, Śukadeva Gosvāmī summarizes his motivation for speaking on this topic as follows:

"In the beginning [the Second and Third Cantos of *Śrīmad-Bhāgavatam*] I have already described how one can progress on the path of liberation. In the *Purāṇas* the vast universal existence, which is like an egg divided into fourteen parts, is described. This vast form is considered the external body of the Lord, created by His energy and qualities. It is generally called the *virāṭ-rūpa*. If one reads the description of this external form of the Lord with great faith, or if one hears about it or explains it to others to propagate *bhāgavata-dharma*, or Kṛṣṇa consciousness, his faith and devotion in spiritual consciousness, Kṛṣṇa consciousness, will gradually increase. Although developing this consciousness is very difficult, by this process one can purify himself and gradually come to an awareness of the Supreme Absolute Truth.

"One who is interested in liberation, who accepts the path of liberation and is not attracted to the path of conditional life, is called *yati*, or a devotee. Such a person should first control his mind by thinking of the *virāṭ-rūpa*, the gigantic universal form of the Lord, and then gradually think of the spiritual form of Kṛṣṇa [*sac-cid-ānanda-vigraha* (BS 5.1)] after hearing of both forms. Thus one's mind is fixed in *samādhi*. By devotional service one can then realize the spiritual form of the Lord, which is the destination of devotees. Thus his life becomes successful.

"My dear King, I have now described for you this planet earth, other planetary systems, and their lands [*varṣas*], rivers and mountains. I have also described the sky, the oceans, the lower planetary systems, the directions, the hellish planetary systems and the stars. These constitute the *virāṭ-rūpa*, the gigantic material form of the Lord, on which all living entities repose. Thus I have explained the wonderful expanse of the external body of the Lord."

25 Prabhupāda's wrote in his purport to SB 1.2.4, "*Śrīmad-Bhāgavatam* is the spotless *Purāṇa*, and it is especially meant for those who desire to get out of the material entanglement permanently."

26 In Prabhupāda's purports to these verses, references are given to the *Garuḍa Purāṇa* and *Śrīmad-Bhāgavatam* 1.3.41.

27 From *Kṛṣṇa: The Supreme Personality of Godhead, A Summary Study of Srila Vyasadeva's Srimad-Bhagavatam,* Tenth Canto, Chapter 89 "The Superexcellent Power of Kṛṣṇa": "Seated on His chariot with Arjuna, Kṛṣṇa proceeded north, crossing over many planetary systems. These are described in *Śrīmad-Bhāgavatam* as *sapta-dvīpa*. *Dvīpa* means 'island.' These planets are sometimes described in the Vedic literature as *dvīpas*. The planet on which we are living is called Jambūdvīpa. Outer space is taken as a great ocean of air, and within that great ocean of air there are many islands, which are the different planets."

28 Please see Appendix 3 for more information about Prabhupāda's quotations on this subject.

29 Letter from A. C. Bhaktivedanta Swami to Svarupa Damodara Das, April 27, 1976, Bhaktivedanta Vedabase, accessed September 10, 2022, https://vedabase.io/en/library/letters/letter-to-svarupa-damodara-15/.

30 For two examples involving Neil deGrasse Tyson and Stephen Hawking engaging this topic in public discourse, see: "Neil DeGrasse Tyson Gets into a Rap Battle With B.o.B Over Flat Earth Theory," January 26, 2016, http://www.npr.org/sections/thetwo-way/2016/01/26/464474518/neil-degrasse-tyson-gets-into-a-rap-battle-with-b-o-b-over-flat-earth-theory, and "Where Are We? Ch. 1 The Circumference of the Earth | Genius by Stephen Hawking," https://florida.pbslearningmedia.org/resource/hawking_genius_ep06_clip01/where-are-we-ch-1-the-circumference-of-the-earth-genius-by-stephen-hawking/, accessed April 1, 2022.

31 Ravi M. Gupta conversation with Krisztina Danka, January 19, 2019, Gainesville, Florida. This recorded interview took place at the "Consciousness in Science" Conference hosted by the Bhaktivedanta Institute for Higher Studies and the UF Center for the Study of Hindu Traditions (CHITRA), January 18-20, 2019, held at the Harn Museum, University of Florida, and the Thomas Center in Gainesville. Conversation transcripts and video recordings for the conference are held by the Bhaktivedanta Institute for Higher Studies, 1515 NW 7th Place, Gainesville, Florida, 32603. For more information please see: bihstudies.org and bit.ly/consci19.

Chapter 1: Guru – Śrīla Prabhupāda on Fifth Canto Purāṇic Cosmography

1 Lecture, *Śrīmad-Bhāgavatam* 3.28.21, Nairobi, November 1, 1975. Prabhupāda had been discussing: "If a material thing like the sun globe ... has got so much ... effulgence that it is keeping lighted all over the universe, just imagine what is the bodily brilliance of Kṛṣṇa ... His bodily light is being reflected in millions and trillions of universes."

2 Lecture, *Bhagavad-gītā* 1.43, London, July 30, 1973.

3 Lecture, *Bhagavad-gītā* 2.12, New York, March 9, 1966.

4 Philosophy Discussion on Charles Darwin with Shyamasundara, c.1973.

5 Lecture (Address), London, September 26, 1969.

6 Lecture, *Bhagavad-gītā* 16.8, Tokyo, January 28, 1975.

7 Lecture, *Bhagavad-gītā* 7.5, Bombay, February 20, 1974.

8 Lecture, *Bhagavad-gītā* 16.8, Tokyo, January 28, 1975.

9 Lecture, *Bhagavad-gītā* 16.8, Tokyo, January 28, 1975.

10 Lecture, *Bhagavad-gītā* 16.8, Tokyo, January 28, 1975.

11 Lecture, *Śrīmad-Bhāgavatam* 1.16.17, Los Angeles, January 12, 1974.

12 Morning walk, L.A., December 9, 1973

13 Letter from A. C. Bhaktivedanta Swami to Dr. Nagendra Babu, Bhaktivedanta Vedabase, accessed September 10, 2022, https://vedabase.io/en/library/letters/letter-to-dr-nagendra-babu/.

14 Of note, while the earth displayed as a polar projection map *would* have a circumference along its perimeter, Prabhupāda was not proposing that his Godbrothers travel the circular perimeter of a disk. By 1969, Prabhupāda had traveled around the world as an airline passenger.

15 This assessment is based on "an in-house study" commissioned by the Bhaktivedanta Book Trust during the 1990s. See: Tamal Krishna Goswami, *A Living Theology of Krishna Bhakti: Essential Teachings of A.C. Bhaktivedanta Swami Prabhupāda*, ed. Graham M. Schweig, (New York: Oxford University Press, 2012), 113.

16 Ferdinando Sardella, *Modern Hindu Personalism: The History, Life and Thought of Bhaktisiddhanta Sarasvati* (New York: Oxford University Press, 2013), 65.

17 Sardella, 66.

18 Sardella, 68–69. Sardella offers an analysis of this account in footnote 47, page 69. He references as a source: "Sambidananda Das, 'The History and Literature of Gaudiya Vaisnavas and their Relation to Other Medieval Vaisnava Schools.' Ph.D. diss., University of London, 1935."

19 Sardella, *Modern Hindu Personalism*, 69. Bhakti Vikāsa Swami notes that without "compromising the principles of traditional *jyotiṣa*," Bhaktisiddhānta appeared "to a certain extent" to incorporate "Western methodology ... particularly by employing modern ... mathematics." See Bhakti Vikāsa Swami, p. 16, fn ‡.

20 Richard L. Thompson, *Vedic Cosmography and Astronomy* (LA: Bhaktivedanta Book Trust, 1989), 8.

21 Quoted in: Bhakti Vikāsa Swami, *Śrī Bhaktisiddhānta Vaibhava*, Volume One (Surat, India: Bhakti Vikas Trust, 2009), 14.

22 Quoted in: Bhakti Vikāsa Swami, 14.

23 Quoted in: Bhakti Vikāsa Swami, 15. Bhaktisiddhānta appeared to be aiming his remarks toward "a new brand of worshipers of *siddhānta*," who "although inexperienced and incomplete in their learning," desired "to attain honor ... in promoting newfangled theories." He noted, "I have more to say about this, which will come in another essay."

24 Sardella, 72.

25 Srimad Bhakti Pradip Tirtha, *Srila Sarasvati Thakur*, 2nd ed. (Calcutta: Gaudiya Mission, 1978), 3. See also Thompson, *Vedic Cosmography*, p. 7, and Bhakti Vikāsa Swami, p. 37.

26 Srimad Bhakti Pradip Tirtha, 73.

27 Bhaktisiddhānta recorded this date in an autobiographical account as the year he received initiation. See Sardella, 80.

28 Sardella, 79.

29 Bhakti Vikāsa Swami, xliv, fn ‡.

30 Sardella, 79.

31 Quoted in Bhakti Vikāsa Swami, 29.

32 Quoted in Bhakti Vikāsa Swami, 25.

33 Quoted in Bhakti Vikāsa Swami, 25–26.

34 Sardella, 69. Sardella also reports that Bhaktisiddhānta "produced Gregorian calendars as well." See p. 70, fn 52: Bhaktisiddhānta Sarasvatī, *Almanac for the year 1917* (Calcutta: Saraswat Chatuspathi, n.d.). Of note, Saraswata Chatuspathi is the name of Bhaktisiddhānta's *jyotiṣa* academy. If produced during years just prior to 1917, this would otherwise be a time focused on Gaudiya Vaishnava missionary activities. In 1918, Bhaktisiddhānta Sarasvatī entered the *sannyāsa* ashram and began establishing centers that would become known as the Gaudiya Math.

35 Thompson, *Vedic Cosmography*, 8.

36 Tamal Krishna Goswami, 113.

37 Bhakti Vikāsa Swami, 297–98.

38 Letter from A. C. Bhaktivedanta Swami to Svarupa Damodara Das, April 27, 1976, Bhaktivedanta Vedabase, accessed September 10, 2022, https://vedabase.io/en/library/letters/letter-to-svarupa-damodara-15/.

39 "Perspectives on Bhaktivedanta Institute," (Philadelphia: Bhaktivedanta Institute, 1979), p. 13.

40 Letter from Richard L. Thompson to Richard W. Sims, May 18, 1978. Box 1, Folder 02, Richard L. Thompson Archives, Alachua, FL, 32615. Thompson's *Vedic Cosmography and Astronomy*, BBT 1989, offers a comprehensive introductory overview of his analysis of the Fifth Canto of the *Śrīmad-Bhāgavatam*. See also Appendix 2, which offers an excerpt taken from the book, specifically discussing the Bhū-maṇḍala plane.

Chapter 2: *Sādhu* – Insights from Previous *Ācāryas*

1 The Sanskrit term offered in the BBT edition of *Śrīmad-Bhāgavatam* verse 5.25.12, as well as 3.11.41: "*aṇu-vat*—just like an atom." In recent times atoms have been considered as displaying a generally spherical form. Meanwhile, ancient concepts similarly placed atomic structure at a foundational level of three-dimensional physicality, albeit in a variety of shapes.

2 Jonathan Edelmann, "Vamshi Dhara Sharman's *Bhāvârthadīpikāprakāśa* and the Original-Reflection Theory," *Journal of Vaishnava Studies* (v. 29, no. 1), 166.

3 A. C. Bhaktivedanta Swami to Sumati Morarjee, August 9, 1976.

4 Richard L. Thompson, "Appendix 1: Vaṁśīdhara on Bhū-maṇḍala and the Earth Globe" in *Vedic Cosmography and Astronomy*, (LA: Bhaktivedanta Book Trust, 1989), 171–180, which offers a thorough analysis of Vaṁśīdhara's commentary. As part of his research, Thompson solicited a full translation of Vaṁśīdhara's commentary on *Srimad-Bhāgavatam* 5.20.38 primarily from two well regarded Sanskrit scholars, V. N. Jha from University of Pune, and V. V. Gangal from the Ananthacharya Indological Research Institute. See: Box 5, Folder 02, Richard L. Thompson Archives, Alachua, FL 32615.

5 See Appendix 4 for a historiographical analysis of Vaṁśīdhara and his commentary.

6 Quoted in Edelmann, "Vamshi Dhara Sharman's *Bhāvârthadīpikāprakāśa*," 168–69.

7 Thompson, *Vedic Cosmography*, 171.

8 Thompson, *Vedic Cosmography*, 172.

9 Quoted in Thompson, *Vedic Cosmography and Astronomy*, 174.

10 Thompson, *Vedic Cosmography*, 174, 176.

11 Thompson, *Vedic Cosmography*, 179.

12 Thompson, *Vedic Cosmography*, 177.

Chapter 3: *Śāstra* – Digging Deeper

1 While the context for the *Bhāgavatam*'s use of these terms could be ambiguous when attempting to distinguish between a description of a spherical object or a round circle, nonetheless "there are no known Purāṇic passages that explicitly state that the earth is flat." See: Toke Lindegaard Knudsen, "Three Purāṇic Statements on the Shape of the Earth," *History of Science in South Asia* 9 (2021): 130. Knudsen's essay analyzes Jñānarāja critique presented in his *Siddhāntasundara* 1.1.27–28, of three arguments that Jñānarāja identified having been attributed to Purāṇic sources. Of the three, Jñānarāja argued (1) and (2) were relevant, and (3) was false. The three perspectives were (1) that the Sanskrit term *bhūgola* referred to an "earth-sphere," (2) Mt. Meru is described as located to the north of everywhere, hence on the north pole of a spherical earth planet, (3) the earth resembling a flat surface of a mirror. While conceding that Jñānarāja's arguments that the Purāṇas clearly identify the earth as a spherical planet proved influential, Knudsen's examination concluded that (1) and (2) where inconclusive as per traditional commentaries on Purāṇic literature, and that rather, these arguments appear reflective of meaning taken from the astronomical Siddhantas. Meanwhile, Knudson was not able to locate a Purāṇic text he could identify with a verse as described in (3), that the earth was shaped like the surface of a flat mirror. Nonetheless, the idea that such a verse existed had been a prevalent supposition.

2 Monier Monier-Williams, Ernst Leumann, and Carl Cappeller, *A Sanskrit-English dictionary etymologically and philologically arranged with special reference to cognate Indo-European languages* (Oxford, The Clarendon Press, 1899), p. 761.

3 Hridayananda das Goswami to Prishni Sutton email, July 14, 2021.

4 Vaṁśīdhara Paṇḍita would also reference this verse in his commentary to SB 5.20.38. See Appendix A.6.

5 Richard L. Thompson, "Exact Science in the *Śrīmad-Bhāgavatam*," *Back to Godhead* 31, no. 4 (July/August 1997): 39–45. See also *Mysteries of the Sacred Universe* for a similar account, Section 4.5, "The Length of the Yojana," 108–17.

6 Quotation from Śrī Bimala Prasad Siddhānta Sarasvatī, *The*

Sūrya-siddhānta and Siddhānta-śiromaṇi, compiled and edited by Danavir Goswami, xx.

7 Letter from A. C. Bhaktivedanta Swami to Madhava, October 3, 1976, Bhaktivedanta Vedabase, accessed September 10, 2021, https://vedabase.io/en/library/letters/letter-to-madhava-4/.

8 Please see Appendix A.4.

9 *The Sūrya-siddhānta and Siddhānta-śiromaṇi*, 20.

10 Richard Thompson, "Planetary Diameters in the *Sūrya-siddhānta*," (Badger, CA: Bhaktivedanta Institute, 1993), 1–3. In the published version, Thompson writes: "Several scholars have argued that the *yojana* in the *Surya-siddhanta* is about 5 miles, thereby bringing the earth's diameter to the realistic value of 5×1600 = 8,000 miles. Examples are Sarma (1956), Burgess (1989), and Dikshit (1969)." See: Richard Thompson, "Planetary Diameters in the *Surya-Siddhanta*," *Journal of Scientific Exploration* 11, no. 2 (1997), 195.

Chapter 4: The Legacy of *Jyotir Veda*: The Tradition Speaks for Itself

1 Bhānu Swami, "Foreword," in *Commentaries on 5th Canto Bhāgavatam Cosmology* by Vaiṣṇava Ācāryas, ed. Danavir Goswami, (Kansas City, MO: Rupanuga Vedic College, 2007), v.

2 Richard L. Thompson, "Vedic Cosmology Seminar 1," ISKCON Philadelphia, July 1, 1986. https://richardlthompson.com/vedic-cosmology-seminar-1

3 Richard L. Thompson, "Astronomy Seminar 1," VIHE Seminar: Gita Nagari, 1990. https://richardlthompson.com/astronomy-seminar-1

4 "Discussion about Bhū-maṇḍala," July 5, 1977, Vṛndāvana, India (770705 - Conversation B). See: https://vanisource.org/wiki/770705_-_Conversation_B_-_Vrndavana

5 Jonathan B. Edelmann, *Hindu Theology and Biology: The Bhāgavata Purāṇa and Contemporary Theory* (Oxford: Oxford University Press, 2012), 177.

6 Edelmann, *Hindu Theology and Biology*, 178.

7 Edelmann, *Hindu Theology and Biology*, 179.

8 Minkowski, "Competing Cosmologies," 351.

9 Minkowski, "Competing Cosmologies," 351.

10 Minkowski, "Competing Cosmologies," 352.

11 For example, in 1503 Jñānarāja authored the *Siddhāntasundara*, the first major *jyotiṣa siddhānta* treatise published since Bhāskarācārya's *Siddhānta-śiromaṇi* in mid-12th century.

12 *Commentaries on 5th Canto Bhagavatam Cosmology* by Vaiṣṇava Ācāryas, ed. Danavir Goswami (Kansas City, MO: Rupanuga Vedic College, 2007), 14.

13 Translation by T. S. Kuppanna Sastry. See: *Vedāṅga Jyotiṣa of Lagandha in its Rk and Yajus Recensions: With the Translation and Notes of Prof. T. S. Kuppanna Sastry*, ed. K. V. Sarma (New Delhi: Indian National Science Academy, 1985), 35.

14 Translation by David Pingree, "The Purāṇas and Jyotihśāstra," 278.

15 N. Chidambaram Iyer, translator. *The Brihad Samhita of Varaha Mihira* (Madura: Printed at South Indian Press, 1884), 11.

Chapter 5: Historical Introduction to India's *Jyotiṣa* Tradition

1 Radha Charan Gupta, "Āryabhaṭa," in *Encyclopaedia of the History of Science, Technology, and Medicine in Non-Western Cultures*, ed. H. Selin (Dordrecht: Springer, 2008).

2 Quoted in S. M. Razaullah Ansari, "Āryabhaṭa I, His Life and His Contributions," *Bulletin of the Astronomical Society of India* 5 (March 1977), 10.

3 Excerpt from Kripa Shankar Shukla and K. V. Sarma, translators, *Āryabhaṭīya of Āryabhaṭa: Critically edited with Introduction, English Translation, Notes, Comments and Indexes* (New Delhi: Indian National Science Academy, 1976), 163.

4 Excerpt from Walter Eugene Clark, translator, *The Āryabhaṭīya of Āryabhaṭa: An Ancient Indian Work on Mathematics and Astronomy* (Chicago: University of Chicago Press, 1930), 81.

5 Shankar and Sarma, translators, *Āryabhaṭīya of Āryabhaṭa*, 118.

6 *Brāhma-sphuṭa Siddhānta, with Vāsanā, Vijñāna and Hindi Commentaries* Vol. I, ed. Ram Swarupa Sharma, (New Delhi: Indian Institute of Astronomical and Sanskrit Research, 1966), 8.

7 R. K. Bhattacharyya, "Brahmagupta: The Ancient Indian Mathematician," in *Ancient Indian Leaps into Mathematics*," eds. B. S. Yadav and M. Mohan (Boston: Brikhäuser, 2011), 185, 187.

8 George Sarton, *Introduction to the History of Science* (Baltimore: Williams and Wilkins, 1947). Quoted in Radha Charan Gupta, "Brahmagupta," in *Encyclopaedia of the History of Science, Technology, and Medicine in Non-Western Cultures*, ed. H. Selin (Dordrecht: Springer, 2008).

9 Bhattacharyya, 186. English translation for the title, "corrected *siddhānta* of Brahma."

10 Translation quoted in Setsuro Ikeyama, "*Brāhmasphuṭasiddhānta* (ch. 21) of Brahmagupta with Commentary of Pṛthūdaka, critically ed. with Eng. tr. and Notes," *Indian Journal of History of Science* 38, no. 3 (New Delhi: Indian National Science Academy, 2003), S154.

11 Bhattacharyya, 186.

12 Ikeyama, S156.

13 Bhattacharyya, 185.

14 Ikeyama, S157.

15 Sharma, 1966, 91.

16 Ikeyama, S164.

17 T. K. Puttaswamy, *Mathematical Achievements of Pre-modern Indian Mathematicians* (London: Elsevier Science, 2012), 142.

18 Ikeyama, S164.

19 Ikeyama, S234–35.

Chapter 6: Bhāskarācārya (c. 1114–1185 CE), with Commentary from Bhaktisiddhānta Sarasvatī Ṭhākura (1874–1937)

1 K. Ramasubramanian, Takao Hayashi, and Clemency Montelle, "Introduction," *Bhāskara-prabhā*, ed. K. Ramasubramanian, Takao Hayashi, Clemency Montelle (Singapore: Springer, 2019), xvii.

2 Plofker, *Mathematics in India*, 173.

3 During this period, Bhaktisiddhānta Sarasvatī would formally publish with the name Siddhānta Sarasvatī. Though some accounts suggest that he used Bhaktisiddhānta Sarasvatī prior, it is generally acknowledged that more common use occurred after the turn of the century, post publication of his *jyotiṣa* work. See: Bhakti Vikāśa Swami, *Śrī Bhaktisiddhānta Vaibhava*, Volume One, (Surat, India: Bhakti Vikas Trust, 2009), xliv, fn. ‡.

4 Quotation from *The Sūrya-siddhānta and Siddhānta-śiromaṇi ganita'dhyaya*, translated and edited by Sri Bimala Prasada Siddhānta

Sarasvatī, compiled and edited by Danavir Goswami (Kansas City, MO: Rupanuga Vedic College, 2007), 178.

5 *Bhāskara-prabhā*, ed. Ramasubramanian, Hayashi, Montelle, 359.

6 *Bhāskara-prabhā*, ed. Ramasubramanian, Hayashi, Montelle, 247. Presented as Verse 1.9 in *Translation of the Sūrya Siddhānta, and of the Siddhānta Śiromaṇi*, translated by Bāpu Deva Śāstrī and Lancelot Wilkinson (1861).

7 K. Ramasubramanian, Takao Hayashi, and Clemency Montelle, "Caṅgadeva's inscription of 1207," in *Bhāskara-prabhā*, eds. K Ramasubramanian Takao Hayashi, Clemency Montelle (Singapore: Springer, 2019), 14.

8 Quoted in Ramasubramanian, Hayashi, Montelle, "Caṅgadeva's inscription," 15. These quotations were originally translated by F. Kielhorn, "Patna inscriptions of the time of Yadava Simghana and his feudatories Soïdeva and Hemadideva," *Epigraphia Indica* 1 (1892), 338–346.

9 Quoted in Ramasubramanian, Hayashi, Montelle, "Caṅgadeva's inscription," 21.

10 An earlier translation phrased the translation, that he had "obtained the favour of the feet of the enemy of Kamsa (Kṛṣṇa)." See: Bhâu Dâjî, "Brief Notes on the Age and Authenticity of the Works of Âryabhaṭa, Varâhamihira, Brahmagupta, Bhaṭṭotpala, and Bhâskarâchârya," *The Journal of the Asiatic Society of Great Britain and Ireland*, New Series 1, no. 1/2 (1865), 418.

11 M. S. Sriram, "*Grahagaṇitādhyāya* of Bhāskarācārya's *Siddhāntaśiromaṇi*," in K. Ramasubramanian et al. *Bhāskara-prabha*, 197.

12 Plofker, *Mathematics in India*, 173, 112.

13 Commentary from Siddhānta Sarasvatī, *The Sūrya-siddhānta and Siddhānta-śiromaṇi*, 327.

14 Quotation from Bāpu Deva Śāstrī and Lancelot Wilkinson, *Translation of the Sūrya Siddhānta, and of the Siddhānta Śiromani.* (Calcutta: Baptist Missionary Press, 1861), 112. See also: Danavir Goswami, *Vedic Cosmology* (Kansas City, MO: Rupanuga Vedic College, 2003), 417.

15 Quotation from Bāpu Deva Śāstrī and Lancelot Wilkinson, *Translation of the Sūrya Siddhānta* 114. See also: Danavir Goswami, *Vedic Cosmology*, 419.

16 Translation from Siddhānta Sarasvatī, *The Sūrya-siddhānta and Siddhānta-śiromaṇi*, 139, 20.

Chapter 7: Jñānarāja (fl. 1503): Commentary at the Time of Caitanya Mahāprabhu

1 Minkowski, "Competing Cosmologies," 352.

2 In a presentation on the life of Rūpa Gosvāmī, Indradyumna Swami describes: "The scripture *Sādhanā-dīpikā* mentions some of the wonderful skills of Rūpa Gosvāmī. The author writes: He could prepare deities, draw expertly, he was a master of astrology and mathematics, [and] he was fluent in Sanskrit, Persian and Bengali, and other languages. He was also a renowned Gauḍīya Vaiṣṇava philosopher and scholar well-known to other *sampradāyas*." See: "Treasures Close to the Heart – Rupa Goswami Part 1," YouTube video, 21:23, posted by "Indradyumna Swami Official," July 31, 2020, https://www.youtube.com/watch?v=JWpbTiFAHzA.

3 *Śrī Caitanya-caritāmṛta*, *Ādi-līlā*, Chapter 13, "The Advent of Lord Śrī Caitanya Mahāprabhu," verse 60 purport, verse 88, verse 90 purport, and verse 121. Nīlāmbara Cakravartī was noted for his accurate calculations pertaining to his grandson's birthchart. See also: CC *Ādi-līlā* 14.13.

On December 29, 1973, during a Morning Walk conversation in Los Angeles, Prabhupāda noted how astronomy, mathematics and astrology were held on near equal footing within traditional culture. ("Astrology is simply based on mathematics ... Astrology is the part of astronomy ... It is calculated by the astral movement. It is a great science. People do not know it now."). At another time Prabhupāda stated: "Astrology is a science. Kṛṣṇa consciousness has nothing to do with astrology, but it is the general custom that as soon as a child is born the astrologers come. That is the Indian system, Vedic system." See: "Conversation on Lord Caitanya Play," Los Angeles, August 4, 1969, https://vanisource.org/wiki/690804_-_Conversation_on_Lord_Caitanya_Play_-_Los_Angeles.

With possible similarity in Western culture, both Galileo Galilei and Johannes Kepler had been respected as professional astrologers during their lifetime. In *Tertius Interveniens* (1610), Kepler described how over time much of the subtlety of the discipline had been lost. Earlier, Kepler had written *De Fundamentis* (1601), whose full title can be translated as "On the More Certain Fundamentals of Astrology."

4 Cintāmaṇi is known for a "lengthy" commentary on his father's work that "was circulated widely in Northern India," which he entitled

Grahagaṇitacintāmaṇi. See: Minkowski, "Competing Cosmologies," 360–64.

As previously noted in the "Introduction," Sūryadāsa concluded one of his major *jyotiṣa* treatises, *Siddhāntasaṃhitāsārasamuccaya*, with a chapter entitled, "Jyotiḥśāstrapurāṇa virodhaparihāra," which translates as the "removal of contradiction(s) of the Purāṇas and Jyotiḥśāstra." Minkowski adds in fn. 57, "There are six sections to Sūryadāsa's viroda chapter, the first untitled, then bhūparidhipramāṇa, ākāravipratipatti, samudradvīpasaṃsthā, grahakakṣākrama, grahakakṣākramasya abhiprāyaḥ." See: Minkowski, "Competing Cosmologies," 365.

5 Toke Lindegaard Knudsen, *The Siddhāntasundara of Jñānarāja: An English Translation with Commentary*. (Baltimore: Johns Hopkins Press, 2014), 5.

6 Minkowski, "On Suryadasa and the Invention of Bidirectional Poetry," 325, 326.

7 Minkowski, "Competing Cosmologies," 355.

8 Christopher Minkowski, "Astronomers and Their Reasons: Working Paper on Jyotihśāstra," *Journal of Indian Philosophy* 30, no. 5 (2002): 497–98.

9 *Siddhāntasundara of Jñānarāja*, 21. Knudsen adds, "Jñānarāja never rejects cosmographical ideas from the Purāṇas. Moreover, he does not hesitate to reject tenets from the Siddhāntas in order to preserve the validity of statements from the Purāṇas." See Knudsen, "Three Purāṇic Statements," 133–34.

10 Minkowski, "Competing Cosmologies," 355.

11 Minkowski, "Competing Cosmologies," 366.

12 Minkowski, "Competing Cosmologies," 356.

13 Minkowski, "The Pandit as Public Intellectual," 83–84.

14 *Siddhāntasundara of Jñānarāja*, 52.

15 *Siddhāntasundara of Jñānarāja*, 54

16 Minkowski, "The Pandit as Public Intellectual," 84.

Knudsen writes in a recent essay along similar lines, "At the time of Jñānarāja, the *Bhāgavatapurāṇa* was growing in popularity, and the number of commentaries was increasing fast." See: Knudsen, "Three Purāṇic Statements," 159.

17 Of note, Jñānarāja's *Siddhāntasundara* preceded the publication of Copernicus's *De revolutionibus orbium coelestium* by several decades.

18 Minkowski, "Competing Cosmologies," 356. Minkowski is primarily referencing verses 1.23 and 1.27 from Jñānarāja's *Siddhāntasundara.*

19 *Siddhāntasundara of Jñānarāja*, 44.

20 *Siddhāntasundara of Jñānarāja*, 105

21 *Siddhāntasundara of Jñānarāja*, 111.

Chapter 8: The *Jyotiṣa* Legacy: Concluding Thoughts

1 Minkowski notes that "Kevalarāma's work was expanded and revised by Nandarāma Miśra in a text entitled *Golodarpaṇa*, completed in 1763." He added that Kevalarāma and Nandarāma Miśra's "arguments are similar enough that I will omit reference to [Miśra's] work here." See: Minkowski, "Competing Cosmologies," 379, fn. 114. Nandarāma Miśra text has also been identified as *Bhūgolakhagola virodhaparihāra.* See: Minkowksi, "The Pandit as Public Intellectual," 83, fn 10. Hopefully future scholarship will bring to light more details on these and similar efforts situated within the historical tradition in a manner readily accessible to a contemporary audience.

2 Thompson, *Vedic Cosmography*, 3. The term Paurāṇikas refers to the followers of Puranic literature.

3 Thompson, *Vedic Cosmography*, 70.

4 Thompson, *Vedic Cosmography*, 3. As mentioned in the "Introduction," there were highly accomplished *jyotiṣa paṇḍitas* within this tradition such as Sūryadāsa during the 1500s and Kevalarāma in the early 1700s who wrote treatises specifically dedicated to the issue of *virodha-parihāra* as an effort to reconcile apparent contradictions when considered from standard human perception. Minkowski notes that "Kevalarāma's work was expanded and revised by Nandarāma Miśra in a text entitled *Golodarpaṇa*, completed in 1763." He noted that Kevalarāma and Nandarama Misra's "arguments are similar enough that I will omit reference to this work here." See: Minkowski, "Competing Cosmologies," 379, fn. 114. Hopefully future scholarship may bring out more details of these past efforts in a manner readily accessible to a contemporary audience.

Chapter 9: Models of Bhū-maṇḍala Active Within ISKCON Discourse

1 Minkowski, "Competing Cosmologies," 352, 378.

2 Of special interest on this point: Minkowski notes that in

Siddhāntasaṃhitāsārasamuccaya (SSSS), Sūryadāsa (1507–1588) may have been the first to consider mathematical incongruities pertaining to the Siddhāntic "standard accommodation" model discussed in the works of prominent *jyotiṣa* scholars such as Lallācārya and Bhāskarācārya. Minkowsi writes: "On the annular oceans and continents Sūryadāsa discovers a problem in the old Siddhāntic accommodation: Although the oceans and continents are placed in the same order in both models, their sizes must be different, as the oceans farther down in the southern hemisphere must be smaller, given the smaller circumference of the earth closer to the poles (SSSS f. 46v, l. 10-f. 47r, l. 1)." See: Minkowski, "Competing Cosmologies," 366, fn. 61.

3 As noted in the "Introduction," Kevalarāma Jyotiṣarāya, the chief astronomer in the court of Mahārāja Jai Singh (1688–1743), appeared to consider a model suggesting some level of correspondence between various models such as proposed by Thompson. Another well-established scholar from within the tradition, Nandarāma Miśra, is said to have "expanded and revised" Kevalarāma's work a few decades later.

In Kevalarāma treatise, *Bhāgavatajyotiṣayor Bhūgola virodhaparihāra*, he wrote of the "reality" that there are "two different earths, both the large, flat Purāṇic earth and also a small, spherical earth floating above the Purāṇic one. The spherical earth is surrounded by the planets, whose orbits are the large ones posited in the Siddhāntas. The planets, therefore, must pass through the surface of the flat Purāṇic earth as they revolve." See Minkowski, "Competing Cosmologies," 379.

Kevalarāma's idea of the geocentric orbits of the observable planets having some form of spatial correspondence with the rings of the massive Purāṇic Bhū-maṇḍala plane if considered from a Euclidean geometric context, could sound reminiscient of Thompson's analysis proposing a comparitably precise mathematical correspondence between the observable motion of the planets with the Purāṇic dimensions given for Bhū-maṇḍala.

4 Letter from A. C. Bhaktivedanta Swami Prabhupāda to S. L. Dhani, Deputy Secretary, November 14, 1976, Bhaktivedanta Vedabase, accessed September 10, 2021, https://vedabase.io/en/library/letters/letter-to-sl-dhani-deputy-secretary/.

5 Richard L. Thompson, *Mysteries of the Sacred Universe: The Cosmology of the Bhāgavata Purāṇa* (Alachua, FL: Govardhan Hill, 2000), 96.

6 The author further describes his work, "We have presented in our *Bhāgavata Cosmology* book, that Bhārata-varṣa consists of nine divisions (*khaṇḍas*). And the ninth portion of Bhārata-varṣa, called

Bhārata-khaṇḍa, is our earth globe. This ninth division of Bhārata-varṣa is known as Bhārata-khaṇḍa or Sudarśana-dvīpa or Bharata-dvīpa or Kumārikā-dvīpa in various *Purāṇas* and the *Mahābhārata*. This can be compared with our earth globe mathematically, philosophically, historically and geographically. This [is] also comparable with the modern observations." Santhoju Pandu to Prishni Sutton email, July 26, 2021.

7 See: R. Caimmi, "Advanced Astronomy in Bhagavata Purana: the Role of the Sun and a Tale from 108," *Applied Mathematical Sciences* 12, no. 5 (2018), 229–254.

8 Giorgio De Santillana and Hertha von Dechend, *Hamlet's Mill: An Essay Investigating the Origins of Human Knowledge and Its Transmission Through Myth* (Boston: Godine, 1969), 61–62.

9 De Santillana and von Dechend, 61–62.

10 Thompson, *Sacred Universe,* 86.

Chapter 10: Empirical Evidence and Natural Science within a Western Context

1 Full title: *Zetetic Astronomy: Earth Not A Globe! An Experimental Inquiry into the True Figure of the Earth Proving it to be an Immovable Plane.* Later editions published in 1873 and 1881 expanded upon the presentation.

2 For a comprehensive historical overview of flat earth perspectives with a focus on recent evangelical initiatives, see: Christine Garwood, *Flat Earth: The History of an Infamous Idea* (New York: Thomas Dunne Books, 2008).

3 Aristotle's arguments for a spherical earth can be found toward the end of Chapter 14 in Book II of his work, *On The Heavens,* credited as having been written 350 BCE. Ptolemy's *Almagest,* credited as having been written around 150 CE, is often considered the most influential work on ancient Greek astronomy in Medieval Europe. Ptolemy also described the earth as a sphere in Book 1, Chapter 4, which is titled "That Also the Earth, Taken as A Whole, is Sensibly Spherical."

4 Within the Christian tradition, Eusebius has been referred to as "Father of Church History." Eusebius is also said to have enjoyed the favor of Constantine the Great (c. 272–337), the first Christian Emperor of the Roman Empire.

5 Eusebius of Caesarea, *Eusebii Pamphili Evangelicae praeparationis* libri XV, trans. Edwin Hamilton Gifford (Oxonii [i.e., Oxford]: Typographeo Academico, 1903), 547–48 (510b–511b). Among the many works

attributed to Aristoxenus are the titles *Life of Socrates* and *Life of Plato*. During Aristoxenus' lifetime, a generation or so after Socrates in Athens, this situation would apparently appear a credible event. The passage cited by Eusebius is referenced to Book Seven of Aristocles' *On Philosophy*.

6 As another aside, Chandragupta Mahārāja and Cānakya Paṇḍita were contemporaries of Alexander the Great, another well-known disciple of Aristotle. According to the ancient Roman historians Plutarch and Justin, Chandragupta as a youth met Alexander on the battlefield. See: Hemchandra Raychaudhuri, *Political History of Ancient India* (New Delhi: Oxford University Press, 1997), 237.

7 Megasthenes, Flavius Arrianus, J. W. McCrindle, and E. A. Schwanbeck, *Ancient India as described by Megasthenês and Arrian; being a translation of the fragments of the Indika of Megasthenês collected by Dr. Schwanbeck, and of the first part of the Indika of Arrian* (Calcutta: Thacker, Spink & Co., 1877), 101.

8 *The Comedy of Dante Alighieri The Florentine : Cantica III Paradise* <*Il Paradiso*>, translated by Dorothy L. Sayers and Barbara Reynolds. (New York: Penguin Books, first published 1962), back cover. Of note, a few decades before Dante, St. Thomas Aquinas (1225–1274) had also described the earth as a sphere in his famous work, *Summa Theologica*, Part I, Question 1, Objection 1, Reply 2. For a comprehensive scholarly work on how this topic was considered in Medieval Europe, see Lesley B. Cormack, "Myth 3: That Medieval Christians Taught That The Earth Was Flat," in *Galileo Goes to Jail and Other Myths About Science and Religion*, ed. Ronald L. Numbers (Cambridge, MA: Harvard University Press, 2009).

9 Whereas initial editions were hand copied manuscripts, the introduction of the movable-type printing press in the 15th century facilitated the reproduction of woodcut illustrations in comparatively mass-produced publications.

10 *The Comedy of Dante Alighieri The Florentine : Cantica II Purgatory* <*Il Purgatorio*>, translated by Dorothy L. Sayers (Baltimore, Md: Penguin Books, 1965), 281.

11 Olaf Pedersen, "In Quest of Sacrobosco," *Journal for the History of Astronomy* 16, no. 3 (1985), 182–84.

12 Translation in Lynn Thorndike, *The Sphere of Sacrobosco and Its Commentators* (Chicago: The University of Chicago Press, 1949), 121.

13 Sacrobosco is said to have received his Aristotle, as it were, for

"demonstrating the roundness of the earth from the observation that the mast of a departing ship disappears from sight after the hull, that the stars rise in the east earlier than farther west, and that different stars are seen at different latitudes," through the works of the famed Arabic scholar, al-Farghani (c. 800–870). Al-Farghani's *Elements of Astronomy on the Celestial Motions* describing Ptolemaic astronomy, along with manuscripts written by other "Arabic astronomers and geographers," had been recently translated from Arabic to Latin during the 12th-century and thus accessible to European scholars such as Sacrobosco. See: Jeffery Burton Russell, *Inventing the Flat Earth: Columbus and Modern Historians* (New York: Praeger, 1991), 19.

In particular, Russell's Chapter Two "The Medieval Ball" (13–26), covers this period. Russell reports that along with Dante and Sacrobosco, the famed Christian theologians and philosophers Saint Augustine of Hippo (354–430) and Saint Thomas Aquinas (1225–74), were well-versed in Greek natural philosophy and wrote about evidence for the spherical nature of the earth. See: 19, 21–23.

14 Thorndike, 121.

15 Contemporary historians consider the notion that Columbus proved the earth round a modern myth. They attribute its source to the early American author, Washington Irving, whose partially fictionalized biography, *The Life and Voyages of Christopher Columbus* (1828), established an exceedingly inaccurate romanticized narrative. In the latter half of the 19th century, this myth was further popularized within an erroneous "conflict between religion and science" thesis famously promoted by John William Draper and Andrew Dickson White.

16 Published images are referenced to having appeared in *Review of Popular Astronomy*, July-August 1966, and *Technology Week*, July 11, 1966, p.12.

17 See: "Molniya 1-3, NSSDCA/COSPAR ID: 1966-035-A," NASA Space Science Data Coordinated Archives. Access date, 28 March 2021: https://nssdc.gsfc.nasa.gov/nmc/spacecraft/display.action?id=1966-035A

18 For general information on the cameras used during various Apollo missions, see Phill Parker, "Apollo-11 Hasselblad Cameras," Apollo Lunar Surface Journal. NASA, at: https://history.nasa.gov/alsj/a11/a11-hass.html (accessed 27 February 2021). For a technical index of images identified with the Apollo 17 mission, see: "Apollo 17 Index: 70 mm, 35 mm, and 16 mm Photographs." Mapping Sciences Branch, Johnson Space Center, NASA (May 1974) 88. (Accessed 27 February 2021) https://www.hq.nasa.gov/office/pao/History/alsj/a17/a17.photidx.pdf

19 NASA offers a selection of DSCOVR satellite images for public viewing at: https://epic.gsfc.nasa.gov

20 It is relatively easy locating a variety of small scale projects that include images using internet search terms such as "curvature of the earth." Without endorsing any particular site, the following address describes a project conducted by physics student team members at the University of Leicester's Astronomy and Rocketry society. See: "Students film breathtaking curvature of Earth using high-altitude weather balloon" by University of Leicester, *Phys Org*, January 6, 2017. Access date March 28, 2021: https://phys.org/news/2017-01-students-breathtaking-curvature-earth-high-altitude.html.

21 The "Geographicus Rare Antique Maps" website offers for sale a high resolution download of the "1936 National Geographic View from the Highest Point Above Earth (South Dakota)" special supplement poster. See bottom right of page for link in red text. The website also offers an informative contextual description with a preview image. Access date 28 March 2021: https://www.geographicus.com/P/AntiqueMap/WorldHighestPoint-natgeo-1936.

22 See "Cabin, 'Explorer II' – 'Long Description'", Smithsonian *National Air and Space Museum*. Access date March 28, 2021: https://airand-space.si.edu/collection-objects/cabin-explorer-ii/nasm_A19370060000.

23 David K. Lynch, "Visually discerning the curvature of the Earth," *Applied Optics* 47, no. 34 (December 1, 2008), H39–H43.

24 Air France and British Airways operated Concordes between 1976 and 2003.

25 Lynch, H41.

26 Lynch, H41.

27 Lynch, H42.

28 Lynch, H41. Lynch also considers how photographic lenses sometimes generate "barrel distortion" which creates an artificial sense of curvature on page H40.

29 Lynch, H41.

30 Shu-Chiu Liu, "Learning from History: A Lesson on the Model of the Earth," *The Science Education Review* 5, no. 1 (2006), 8.

31 Liu, 9.

32 Also see: Appendix A.4 "Evidence of a Spherical Earth."

Chapter 11: Conclusion

1 *Śrī Caitanya-caritāmṛta, Madhya-līlā* 6.135, Purport.

2 *Śrīmad-Bhāgavatam* 5.16.4.

3 *Śrīmad-Bhāgavatam* 5.16.4, Purport.

4 A familiarity with basic observable astronomy is also integral to more mundane activities like charting agricultural calendars for planting as well as navigation, particularly when in unfamiliar territory or out of sight of land.

5 *Śrīmad-Bhāgavatam* 5.16.4, Purport.

6 *Śrī Caitanya-caritāmṛta, Adi-lila* 3.8, Purport.

7 *Śrīmad-Bhāgavatam* 1.1.1, Purport.

8 Letter from A. C. Bhaktivedanta Swami to Svarupa Damodara Das, April 27, 1976, Bhaktivedanta Vedbase, accessed September 10, 2021, https://vedabase.io/en/library/letters/letter-to-svarupa-damodara-15/.

9 Some of these works also consider age-old questions from within the tradition such as: Do the *Sūrya-siddhānta* and the *Purāṇas* offer a description of the same universe manifest in different *kalpas*? Or do these apparently differing accounts combinedly offer a glimpse into a complex set of metaphysics describing the present cosmos? Could the Purāṇic model depict a multidimensional reality in which standard human perception is limited to three-dimensional geometric space? Would an inability to potentially perceive the full breadth of cosmological phenomena be limited by present karmic qualifications and a conditioned level of sense perception?

10 Considered in this light, perhaps endeavoring to manipulate the material energy, intellectually or otherwise, is not the essential point of the *bhakti* tradition.

11 *Śrīmad-Bhāgavatam* 5.16.4, Purport.

12 Of note, during the 16th century Copernicus famously quantified heliocentric motion while extensively utilizing Greek methodologies, and his model proved hardly more accurate than Ptolemy's at mathematically predicting motion. With the shift toward a heliocentric model culminating in the Newtonian Synthesis occurring during the 17th century, an understanding of the known universe remained focused on what is now described as the solar system, with the stars fixed in the background not far beyond Saturn. In other words, the idea of "universe" encompassed a modern concept of "solar system" for much

of this period. Meanwhile, early 20th-century astronomers considered the Milky Way the central feature of the universe, with a series of gaseous cloud nebulae situated beyond it. The use of the Mt. Wilson Observatory's 100-inch light telescope during the 1920s suggested that these nebulae were actually huge galaxies much like the Milky Way, and expanding at great distance. Meanwhile, more recent astrophysics takes account of a comparatively unlimited expanding universe filled with countless galaxies, with everything presently observable but a small part of all that should be observable, in order to justify celestial motion according to the known laws of physics. The rest, dark matter and dark energy, invites further research. In other words, the observable universe presently accessible to modern science is regarded as but a small percentage of all that should be out there, and a far cry from appreciating the Milky Way as playing a central role. Perhaps by comparison the early modern shift from geocentric to heliocentric perspectives could well prove considerably less dramatic than the implications of recent discoveries in astrophysics.

13 *The Siddhāntasundara of Jñānarāja*, 111.

14 *Bhagavad-gītā* 10.41, Purport.

Appendix 4b: Description of the lineage of the commentator

1 Note from translator: "This statement invites affirmation as to location on a modern map. There could also be a pond or body of water which is known as Rāmkuṇḍa or Rāma-sarovar. I looked closely at a map and found that in Khandara there is a pond known as Govinda Sāgar, but nothing related to Rāma."

2 This date would be 1888 according to the Gregorian calendar.

Appendix 4c: A historiographical account of biographical material on the life of the Vaishnava commentator, Śrī Va□śīdhara

3 Likely also known as Sri Venkateswar Steam Press.

4 Parashuram Krishna Gode, "Date of Navanītarāma's Commentary on the Radhuvaṁśa," *Journal of the Oriental Institute* Vol. III, No. 3 (March 1954), 281.

5 Thompson, *Vedic Cosmography*, 164.

6 V.S. Agarwal, "Foreword," in *Prākṛita-Paṅgalaṁ: A text on Prākṛita and Apabhraṁśa Metres*, Part I, ed. Dr. Bhola Shanker Vyas (Varanasi, India: Prakrit Text Society, 1959), ii.

7 Richard L. Thompson Archives, "Astronomy Seminar 2. VIHE Seminar: Gita Nagari, PA, 1990," http://richardlthompson.com/astronomy-seminar-2, access date September 20, 2021.

8 "Interviews with Professors O'Connell, Motilal and Shivaram; Toronto, June 18, 1976," VaniSource.org, https://vanisource.org/wiki/760618_-_Interview_B_-_Toronto, access date September 21, 2021.

Note: Vanisource.org appears to misspell Bimal Krishna Matilal's last name as Bimal Motilal. During the discourse, Prabhupāda and Matilal speak in Bengali. Matilal was born in West Bengal. He would pursue studies at Sanskrit College (in Calcutta) as well as the University of Calcutta, prior to earning his PhD at Harvard University during the 1960s. See: Purushottama Bilimoria and Jay L. Garfield, "Editorial: Bimal Krishna Matilal, 1935–1991," in *Sophia* 55 (2016), 455–458.

9 Karl Potter, *Encyclopedia of Indian Philosophies*, Vol. 9, *Acintyabhedhābheda Vaiṣṇava Philosophy* (Delhi: Motilal Banarsidass, 2015), 227.

10 Karl Potter, *Bibliography of Indian Philosophies*: "Part I Texts Whose Authors Can Be Dated Authors Listed Chronologically, 15th century to the present," 4/15/2020, https://faculty.washington.edu/kpotter/ckeyt/txt4.htm, access date September 30, 2021. Note: Karl Potter passed away in January 2022, and as of July 2023, his faculty website hosting *Bibliography of Indian Philosophies* is no longer available.

11 Jonathan Edelmann, "Vamshi Dhara Sharman's *Bhāvârthadīpikāprakāśa*," 167.

12 Appendix 4c offers an English translation of Vaṁśīdhara's autobiographical writings.

13 Edelmann, 165.

14 Edelmann, 167.

15 Edelmann, 177.

16 Edelmann, 167.

17 Edelmann, 166. Edelmann is referencing: H. G. Shastri, ed., *The Bhāgavat [Śrīmad Bhāgavata Mahapurāṇa]*, (Ahmedabad: BJ Institute, 1996), XXXVIII.

18 Edelmann, 166.

19 Edelmann, 166.

20 Ludo Rocher, *A History of Indian Literature: The Purāṇas*, second volume (fasc. 3), (Wiesbaden: Otto Harrassowitz, 1986), 149.

21 Ācārya Paṇḍita Śrī Rāmamūrti Śāstrī Paurāṇika, ed., *Śrīmad-bhāgavata-mahāpurāṇam with the Gūḍhārtha-dīpinī Sanskrit Commentary*, Volume 1, Skandhas 1–9 (Varanasi: Chaukhamba Vidyabhavan, 1993), front matter.

22 Ekkehard Lorenz, "The Guru, Mayavadins, and Women: Tracing the Origins of Selected Polemical Statements in the Work of A.C. Bhaktivedanta Swami," in *The Hare Krishna Movement: The Postcharismatic Fate of a Religious Transplant*, eds. Edwin Bryant and Maria Ekstrand (New York: Columbia University Press, 2004), 126, fn. 18.

Appendix 5: Evidence for a Spherical Earth

1 The translation of verses 12.23–41 from the *Sūrya-Siddhānta* offered by the author appears primarily drawn from: Ebenezer Burgess, Phanindralal Gangooly, and Prabodhchandra Sengupta, *Translation of the Sūrya-Siddhānta: a text-book of Hindu astronomy, reprinted from the edition of 1860* (Calcutta: University of Calcutta, 1935).

Appendix 6: Polar Projection Maps and Flat Earth Arguments

1 Matthias Tomczak, "Science, Civilization and Society: Cellarius' Star Atlas," 2004, https://incois.gov.in/Tutor/science+society/lectures/illustrations/lecture20/cellarius.html, access date May 18, 2021.

2 Liz Fuller-Wright, "Princeton astrophysicists re-imagine world map, designing a less distorted, 'radically different' way to see the world," 15 February 2021, https://www.princeton.edu/news/2021/02/15/princeton-astrophysicists-re-imagine-world-map-designing-less-distorted-radically, access date: 18 May 2021. See also: "Flat Maps that improve on the Winkel Tripel," by J. Richard Gott III, David M. Goldberg, and Robert J. Vanderbei. https://arxiv.org/pdf/2102.08176v1.pdf, access date May 18, 2021.

Appendix 7: Sanskrit Mathematics in Western Antiquity

1 Justinian is credited with building the Hagia Sophia (Holy Wisdom), considered one of the world's greatest domed cathedrals. There are also reports that Justinian proved to be an emperor of good taste, as he retained "an Indian cook" in the service of his royal court. See: A. L. Basham, *The Wonder That Was India*. London: Sidgwick and Jackson, 1954. p. 229. This historical factoid helps to illustrate the extent of cross cultural exchange that took place during this period.

2 For an engaging introduction to the Saint Thomas Christians of India, see: Paul Zacharia; photographs by Lynn Johnson, "The Surprising Early History of Christianity," *Smithsonian Magazine, Smithsonian Journeys Travel Quarterly India Issue*, February 19, 2016. www.smithsonianmag.com/travel/how-christianity-came-to-india-kerala-180958117/

3 John M. McMahon, "Severus Sebokht," *The Biographical Encyclopedia of Astronomers, Springer Reference.* New York: Springer, 2007. pp. 1044–45. https://link.springer.com/referenceworkentry/10.1007/978-0-387-30400-7_1260

4 Ighnāṭyūs Afrām I, *The Scattered Pearls: A History of Syriac Literature and Sciences* Second Revised Edition, translated and edited by Matti Moosa. New Jersey: Gorgias Press, 2003.

5 Quoted in A. L. Basham, *The Wonder That Was India.* London: Sidgwick and Jackson, 1954, pg. vi. Basham used this paragraph as the epigraph for his book.

6 Basham wrote: "For long it was thought that the decimal system of numerals was invented by the Arabs, but this is certainly not the case. The Arabs themselves called mathematics 'The Indian (art)' (*hindisat*), and there is now no doubt that the decimal notation, and other mathematical lore, was learnt by the Muslim world either though merchants trading with the west coast of India, or through the Arabs who conquered the Sind in AD 712." Basham also noted, "Medieval Indian mathematicians, such as Brahmagupta (7th century), Mahāvīra (9th century), and Bhāskara (12th century), made several discoveries which in Europe were not known until the Renaissance or later." See: Basham, p. 496.

7 Basham, p. 496. Basham further commented, "Severus Sebokht's 'nine symbols' would be quite useless for expressing quantities over nine without a zero sign and place notation." See: Basham, p. 495 fn†

8 Basham, 495.

9 Quoted in Hidemi Takahashi, "Between Greek and Arabic: The Sciences in Syriac from Severus Sebokht to Barhebraeus," In *Transmission of Sciences: Greek, Syriac, Arabic and Latin*, eds. Kobayashi Haruo and Kato Mizue. Tokyo: Organization for Islamic Areas Studies, Waseda University, 2010. p. 23.

Appendix 8: Personal Testimony: "A View from 41,000 feet"

1 Declan Fahy, author of *New Celebrity Scientists: Out of the Lab and Into the Limelight* (Lanham, MD: Rowman & Littlefield, 2015), includes a

chapter with the title, "The Charming Stardom of Brian Greene." In it he describes Greene as the "cofounder and director of Columbia's Institute for Strings, Cosmology, and Astroparticle Physics, [who] came to public prominence as string theory became the dominant but controversial approach to turn-of-the-century theoretical physics. He also represents a dramatic, unprecedented point in the history of science where at stake is fundamental reconsideration of what constitutes reality, and what counts as a valid scientific theory about the nature of the physical world. He is an eloquent explainer of the bizarre ideas that have captured the public imagination, as black holes did in the late 1970s. Greene also embodies, more so than any of the other scientists profiled here, the modern telegenic scientist whose public status relies on his ability to glide between the worlds of leading-edge research and popular entertainment without sacrificing his scientific authority. Viewed as a whole, his career shows him to be a new breed of public scientist" (p. 158).

2 On page 175 of Fahey's book mentioned in the previous footnote, the author writes concerning Greene's family background and metaphysical views:

Greene—who was born Jewish but is not religious—is aware of the spiritual and metaphysical interpretations of strings. His half-brother, Joshua, is a Hare Krishna devotee, and Greene often told him about new ideas in physics. "And a lot of times, he'll say to me, 'Well, we already knew that. That's in Vedic text 23 or something of that sort,' and, you know," said Greene, "it's an interesting exchange because when we've ever go into more detail on it, it seems as though many of the ideas that we have come upon do have a kind of resonance with ideas that have been articulated in ancient text or even in more modern theological or mystical text" (*Fresh Air*, NPR, February 11, 2005). Elsewhere Greene remarked: "So, from my dad came two seekers. One went the religious route, one the scientific route" (Anjana Ahuja, "He's Got an Answer for Everything," *Times* (London), February 19, 2004, 10).

But Greene deftly sidesteps the inevitable science-and-religion debate. In interviews, he has said he does not see science and religion in conflict, except in some narrow context of teaching creationism instead of evolution. "My own personal view is science can't rule out religion. It can't rule out God. God could be behind it all, set it all up so we've discovered what we've discovered but set it up so that we don't have any direct evidence for his existence or her existence, whatever the right word is," he told *The Daily Beast*. "My view is if what we are doing as scientists is working out God's design, and that's what we're doing, if that's the case, I'm thrilled to be part of that momentous journey. But if that's not the case—and I don't think it is—if what we're doing is just working out the deep laws of the cosmos that brought the universe into existence and have guided its evolution, if I'm part of that journey and I can take that a little bit further, I'm thrilled to be part of that. So, in a day-to-day way, it just doesn't matter" ("Welcome to the Multiverse," *The Daily Beast,* May 21, 2012, accessed June 20, 2014, http://www.briangreene.org/bg_beast.html). [Note: *The Daily Beast* link offered in Fahy's account no longer appears active as of July 2023.]

Bibliography

Afrām I, Ighnāṭyūs. *The Scattered Pearls: A History of Syriac Literature and Sciences*, Revised Second Edition. Translated and edited by Matti Moosa. Piscataway, N.J: Gorgias Press, 2003.

Agarwal, V. S. "Foreword." In *Prākṛita-Paṅgalaṁ: A text on Prākṛita and Apabhraṁśa Metres*, Part I, edited by Dr. Bhola Shanker Vyas, i–iv. Varanasi, India: Prakrit Text Society, 1959.

Ansari, S. M. Razaullah. "Āryabhaṭa I, His Life and His Contributions." *Bulletin of the Astronomical Society of India* 5, no. 1 (March 1977): 10–18.

Āryabhaṭa, and Walter Eugene Clark. *The Āryabhaṭīya of Āryabhaṭa: an ancient Indian work on mathematics and astronomy*. Chicago: University of Chicago Press, 1930

Āryabhaṭa, Kripa Shankar Shukla, and K. V. Sarma. *Āryabhaṭīya of Āryabhaṭa: critically edited with introduction, English translation, notes, comments and indexes*. New Delhi: Indian National Science Academy, 1976.

Basham, A. L. *The Wonder That Was India*. London: Sidgwick and Jackson, 1954.

Bhaktivedanta Swami Prabhupāda, A. C. *Bhagavad-gītā As It Is*, Los Angeles: Bhaktivedanta Book Trust, 1989.

———. *Śrīmad-Bhāgavatam*, Los Angeles: Bhaktivedanta Book Trust, 1987.

———. *Śrī Caitanya-caritāmṛta*, Los Angeles: Bhaktivedanta Book Trust, 1996.

Bhattacharyya, R. K. "Brahmagupta: The Ancient Indian Mathematician." In *Ancient Indian Leaps into Mathematics*," edited by B.S. Yadav and M. Mohan, 185–192. Boston: Brikhäuser, 2011.

Brahmagupta, Ram Swarup Sharma, Pṛthūdakasvāmin, and Sudhākara Dvivedī. *Brāhma-sphuṭa Siddhānta, with Vāsanā, Vijñāna and Hindi Commentaries* Vol. I. Edited by a board of editors headed by Acharyavara Ram Swarup Sharma. New Delhi: Indian Institute of Astronomical and Sanskrit Research, 1966.

Brooke, John Hedley. *Science and Religion: Some Historical Perspectives*. New York: Cambridge University Press, 1991.

Burgess, Ebenezer, Phanindralal Gangooly, and Prabodhchandra Sengupta. *Translation of the Sūrya-Siddhānta: a textbook of Hindu astronomy, reprinted from the edition of 1860*. Calcutta: University of Calcutta, 1935.

Caimmi, R. "Advanced Astronomy in Bhagavata Purana: The Role of the Sun and a Tale from 108." *Applied Mathematical Sciences* 12, no. 5 (2018): 229–254.

Cormack, Lesley B. "Myth 3: That Medieval Christians Taught that the Earth was Flat." In *Galileo Goes to Jail and Other Myths About Science and Religion*, edited by Ronald L. Numbers, 28–34. Cambridge, MA: Harvard University Press, 2009.

Dâjî, Dr. Bhâu. "Brief Notes on the Age and Authenticity of the Works of Âryabhaṭa, Varâhamihira, Brahmagupta, Bhaṭṭotpala, and Bhâskarâchârya." *The Journal of the Asiatic Society of Great Britain and Ireland* New Series 1, no. 1/2 (1865): 392–418.

Dalela, Ashish. *Mystic Universe: An Introduction to Vedic Cosmology*. Bangalore: Shabda Press, 2016.

Dante, Alighieri. *The Comedy of Dante Alighieri The Florentine : Cantica III Paradise <Il Paradiso>*. Translated by Dorothy L. Sayers and Barbara Reynolds. New York: Penguin Books, 1962.

_______. *The Comedy of Dante Alighieri The Florentine : Cantica II Purgatory <Il Purgatorio>*. Translated by Dorothy L. Sayers. Baltimore, Md: Penguin Books, 1965.

Das, Pavaneshwar. *Bhāgavata Cosmology: Vedic Alternative to Modern Cosmology*. Mumbai: Tulsi Books, 2019.

De Santillana, Giorgio, and Hertha von Dechend, *Hamlet's Mill: An Essay on Myth and the Frame of Time*. Boston: David R. Godine, 2002.

Edelmann, Jonathan B. *Hindu Theology and Biology: The Bhagavata Purana and Contemporary Theory*. Oxford: Oxford University Press, 2012.

_______. "Vamshi Dhara Sharman's *Bhāvârthadīpikāprakāśa* and the Original-Reflection Theory." *Journal of Vaishnava Studies* 29 (2020): 165–179.

Eusebius of Caesarae. *Eusebii Pamphili Evangelicae praeparationis* libri XV. Translated by Edwin Hamilton Gifford. Oxonii [i.e. Oxford]: Typrographeo Academico, 1903.

Fahy, Declan. *New Celebrity Scientists: Out of the Lab and into the Limelight*. Lanham, MD: Rowman & Littlefield, 2015.

Garwood, Christine. *Flat Earth: The History of an Infamous Idea*. New York: Thomas Dunne Books, 2008.

Gode, Parashuram Krishna. "Date of Navanītarāma's Commentary on the Radhuvaṁśa." *Journal of the Oriental Institute* III (1954): 277–282.

Goswami, Danavir, ed., A. C. Bhaktivedanta Swami Prabhupāda, Śrīdharasvāmin, et al. *Commentaries on 5th Canto Bhāgavatam Cosmology*. Kansas City, MO: Rupanuga Vedic College, 2007.

Goswami, Danavir, ed., Siddhānta Sarasvatī Bimala Prasada, and Bhāskarācārya. *The Sūrya-siddhānta and Siddhānta-śiromaṇi* (*Graha ganita'dhyaya*). Kansas City, MO: Rupanuga Vedic College, 2007.

Goswami, Tamal Krishna. *A Living Theology of Krishna Bhakti: Essential Teachings of A. C. Bhaktivedanta Swami Prabhupāda*, edited by Graham M. Schweig. New York: Oxford University Press, 2012.

Gupta, R. C. "Āryabhaṭa." In *Encyclopaedia of the History of Science, Technology, and Medicine in Non-Western Cultures*, edited by Helaine Selin. Dordrecht: Springer, 2008.

Hock, Hans Henrich. "Philology and the Historical Interpretation of the Vedic Texts." In *The Indo-Aryan Controversy: Evidence and Inference in Indian History*, edited by Edwin F. Bryant and Laurie L. Patton, pp. 282–308. New York: Routledge, 2005.

Ikeyama, Setsuro. "*Brāhmapsuṭasiddhānta* (ch.21) of Brahmagupta with Commentary of Pṛthūdaka, critically edited with English translation and notes." *Indian Journal of History of Science* 38 (2003): no. 1: S1–S74; no. 2: S75–S152; no. 3: S153–S236; no. 4: S237–S308.

Irving, Washington. *A History of the Life and Voyages of Christopher Columbus*. London: John Murray, 1828.

Iyer, N. Chidambaram, trans. *The Bṛihat Saṃhitâ of Varaha Mihira*. Madura: Printed at South Indian Press, 1884.

Jain, Pushpa Kumari. *The Sūryaprakāśa of Sūryadāsa*, vol. I. Vadodara: Oriental Institute, 2001.

Kielhorn, F., trans. "Patna inscriptions of the time of Yādava Simghana and his feudatories Soïdeva and Hemadideva." *Epigraphia Indica* 1 (1892): 338–346.

Knudsen, Toke Lindegaard. *The Siddhāntasundara of Jñānarāja: An English Translation with Commentary*. Baltimore: Johns Hopkins Press, 2014.

________. "Three Purāṇic Statements on the Shape of the Earth." *History of Science in South Asia* 9 (2021): 128–166.

Lagadha, T. S. Kuppanna Sastry, and K. Venkateswara Sarma. *Vedāṅga Jyotiṣa of Lagadha in Its Ṛk and Yajus Recensions: With the Translation and Notes of Prof. T. S. Kuppanna Sastry*. New Delhi: Indian National Science Academy, 1985.

Lorenz, Ekkehard. "The Guru, Mayavadins, and Women: Tracing the Origins of Selected Polemical Statements in the Work of A.C. Bhaktivedanta Swami." In *The Hare Krishna Movement: The Postcharismatic Fate of a Religious Transplant*, edited by Edwin F. Bryant and Maria L. Ekstradt, 112–128. New York: Columbia University Press, 2004.

Liu, Shu-Chiu. "Learning from History: A Lesson on the Model of the Earth." In *The Science Education Review* 5, no. 1 (2006): 6–13.

Lynch, David K. "Visually discerning the curvature of the Earth." *Applied Optics* 47, no. 34 (2008): H39–H43.

McMahon, John, M. "Severus Sebokht." In *The Biographical Encyclopedia of Astronomers,* edited by Thomas Hockey, et al., 1044–1045. New York: Springer, 2007.

Megasthenês, Flavius Arrianus, J. W. McCrindle, and E. A. Schwanbeck. *Ancient India as described by Megasthenês and Arrian: being a translation of the fragments of the Indika of Megasthenês collected by Dr. Schwanbeck, and of the first part of the Indika of Arrian.* Calcutta: Thacker, Spink & Co., 1877.

Minkowski, Christopher. "The Pandit as Public Intellectual: The Controversy over *virodha* or Inconsistency in the Astronomical Sciences." In *The Pandit: Traditional Scholarship in India*, edited by Axel Michaels, 79–96. New Delhi: Manohar Publishers, 2001.

________. "Astronomers and Their Reasons: Working Paper on Jyotihśāstra." *Journal of Indian Philosophy* 30, no. 5 (2002): 495–514.

________. "On Sūryadāsa and the Invention of Bidirectional Poetry (*vilomakāvya*)." *Journal of the American Oriental Society* 124, no. 2 (2004): pp. 325–333.

________. "Competing Cosmologies in Early Modern Indian Astronomy." In *Studies in the History of the Exact Sciences in Honour of David Pingree*, edited by Charles Burnett, et al., 349–85. Leiden: Brill, 2004.

Monier-Williams, Monier, Ernst Leumann, and Carl Cappeller. *A Sanskrit-English dictionary etymologically and philologically arranged with special reference to cognate Indo-European languages.* Oxford, The Clarendon Press, 1899.

Montelle, Clemency, and Kim Plofker. "Integrating Logarithm Concepts into Syntheses of Sanskrit, Muslim and European Astronomy in 18th-Century Jaipur." In *Science and Society in the Sanskrit World*, edited by Christopher T. Fleming, Toke Lindegaard Knudsen, Anuj Mistra, and Vishal Sharma, pp. 172–217. Leiden: Brill, 2023.

Pedersen, Olaf. "In Quest of Sacrobosco." *Journal for the History of Astronomy* 16, no. 3 (1985): 175–221.

Pingree, David. "The Purāṇas and Jyotiḥśāstra: Astronomy." *Journal of the American Oriental Society* 110, no. 2 (April–June 1990): 274–280.

Plofker, Kim. *Mathematics in India.* Princeton: Princeton University Press, 2009.

Potter, Karl. *Encyclopedia of Indian Philosophies,* Vol. 9, *Acintyabhedhābheda Vaiṣṇava Philosophy.* Delhi: Motilal Banarsidass, 2015.

Puttaswamy, T. K. *Mathematical Achievements of Pre-modern Indian Mathematicians.* London: Elsevier Science, 2012.

Ramasubramanian, K, Takao Hayashi, and Clemency Montelle, eds. *Bhāskara-prabhā: Sources and Studies in the History of Mathematics and Physical Sciences.* Singapore: Springer, 2019.

Raychaudhuri, Hemchandra. *Political History of Ancient India.* New Delhi: Oxford University Press, 1997.

Rocher, Ludo. *A History of Indian Literature: The Purāṇas,* second volume (fasc. 3). Wiesbaden: Otto Harrassowitz, 1986.

Russell, Jeffery Burton. *Inventing the Flat Earth: Columbus and Modern Historians.* New York: Praeger, 1991.

Sarasvatī, Bhaktisiddhānta. *Almanac for the Year 1917.* Calcutta: Saraswat Chatuspathi, n.d.

Sardella, Ferdinando. *Modern Hindu Personalism: The History, Life and Thought of Bhaktisiddhānta Sarasvatī.* New York: Oxford University Press, 2013.

Sarma, K. V., ed. *Vedanga Jyotiṣa of Lagandha in its Rk and Yajus Recensions.* New Delhi: Indian National Science Academy, 1985.

Sarton, George. *Introduction to the History of Science.* Baltimore: Williams and Wilkins, 1947.

Śāstrī, Bāpu Deva, and Lancelot Wilkinson. *Translation of the Sūrya Siddhānta and of the Siddhānta Śiromaṇi.* Calcutta: Baptist Mission Press, 1861.

Śāstrī Paurāṇika, Ācārya Paṇḍita Śrī Rāmamūrti, ed. *Śrīmad-bhāgavata-mahāpurāṇam with the Gūḍhārtha-dīpinī Sanskrit Commentary,* Volume 1, Skandhas 1–9. Varanasi: Chaukhamba Vidyabhavan, 1993.

Sharma, Virendra Nath. "Sawai Jai Singh's Hindu Astronomers." *Indian Journal of History of Science* 28, no. 2 (1993): 137–155.

Sriram, M. S. "*Grahagaṇitādhyaya of Bhaskarācārya's Siddhāntasiromaṇi.*" In *Bhāskara-prabhā: Sources and Studies in the History of Mathematics and Physical Sciences,* edited by K. Ramasubramanian, Takao Hayashi, and Clemency Montelle, 197–231. Singapore: Springer, 2019.

Swami, Bhānu. "Foreword." In *Commentaries on 5th Canto Bhāgavatam Cosmology,* edited by Danavir Goswami. Kansas City, MO: Rupanuga Vedic College, 2007.

Swami, Bhakti Vikāśa. *Śrī Bhaktisiddhānta Vaibhava: The Grandeur and Glory of Śrīla Bhaktisiddhānta Sarasvatī Ṭhākura*, Vol. 1. Surat: Bhakti Vikas Trust, 2009.

Takahashi, Hidemi. "Between Greek and Arabic: The Sciences in Syriac from Severus Sebokht to Barhebraeus." In *Transmission of Sciences: Greek, Syriac, Arabic and Latin*, edited by Haruo Kobayashi and Mizue Kato, 16–22. Tokyo: Organization for Islamic Areas Studies, Waseda University, 2010.

Tirtha, Srimad Bhakti Pradip. *Srila Sarasvati Thakur*, 2nd ed. Calcutta: Gaudiya Mission, 1978.

Thompson, Richard L. "Vedic Cosmology Seminar 1." Recorded July 1, 1986, at ISKCON Philadelphia, PA. Audio, 75:36, with transcript. https://richardlthompson.com/vedic-cosmology-seminar-1.

________. *Vedic Cosmography and Astronomy*. LA: Bhaktivedanta Book Trust, 1989.

________. "VIHE Astronomy Seminar 1." Recorded 1990 at Gita Nagari, Port Royal, PA. Audio, 86:47, with transcript. https://richardlthompson.com/astronomy-seminar-1.

________. "Planetary Diameters in the *Sūrya-siddhānta*." Badger, CA: Bhaktivedanta Institute, 1993.

________. "Planetary Diameters in the *Sūrya-siddhānta*." *Journal of Scientific Exploration* 11, no. 2. (1997): 193–200.

________. *Mysteries of the Sacred Universe: The Cosmology of the Bhāgavata Purāṇa*. Alachua, FL: Govardhan Hill, 2000.

Thorndike, Lynn, and Johannes de Sacrobosco. *The Sphere of Sacrobosco and Its Commentators*. Chicago: University of Chicago Press, 1949.

Glossary of Terms

ācārya – one who teaches by his own example; a spiritual master.

acintya-bhedābheda-tattva – Caitanya Mahāprabhu's doctrine of the "inconceivable oneness and difference" of God and His energies.

Antarikṣa – the narrow space between the upper and lower hemispheres of the Brahmāṇḍa, literally "inner space."

Āryabhaṭa (b. 476 CE) – famed Indian astronomer whose *Āryabhaṭīya* described the earth as rotating on its axis.

astrolabe – an astronomical instrument with stereographic projections of the earth, some stars, and the ecliptic on a series of metal plates.

avatāra – a fully or partially empowered incarnation of God who descends from the spiritual realm for a particular mission.

bhakti – devotional service to the Supreme Lord. Perfected *bhakti,* pure love of God, is described as the ultimate goal within the Vaishnava tradition.

Bhaktisiddhānta Sarasvatī Ṭhākura (1874–1937) – the spiritual master of His Divine Grace A. C. Bhaktivedanta Swami Prabhupāda; an *ācārya* in the Gaudiya Vaishnava disciplic succession who established the Gaudiya Math in India. In the years prior to taking *sannyāsa*, considered a leading Sanskrit mathematician and scholar of *siddhāntic* astronomy.

Bhaktivedanta – a title conferred upon A. C. Bhaktivedanta Swami by the Gaudiya Math, meaning "one who has understood that the conclusion of Vedic scripture is *bhakti* (devotional service)."

Bhaktivedanta Swami, A. C. (1896–1977) – Founder-*Ācārya* of the International Society of Krishna Consciousness, commonly known as the "Hare Krishna Movement." Disciple of Bhaktisiddhānta Sarasvatī Ṭhākura. Known to his followers as Srila Prabhupāda, he is widely regarded as the foremost Vedic scholar, translator, and teacher of the modern era.

Bhaktivinoda Ṭhākura (1838–1914) – an *ācārya* in the Gaudiya Vaishnava disciplic succession; the father of Bhaktisiddhānta Sarasvatī Ṭhākura. Known for envisioning a world-wide society of Vaishnavas,, and providing groundwork toward that end.

Bhāgavata Purāṇa – one of the most famous and popular of the *Purāṇas*, especially among the Vaishnavas, and one which contains an extensive section on cosmology. It is also known as *Śrīmad-Bhāgavatam*, or simply *Bhāgavatam*.

Bhāgavatam – same as *Bhāgavata Purāṇa*.

Bhārata-varṣa – southernmost of the nine subdivisions (*varṣas*) of Jambūdvīpa, said to refer either to India or to the earth as a whole.

Bhāskarācārya (c. 1114–1185 CE) – Indian astronomer who wrote the *Jyotiḥśāstra* known as *Siddhānta-śiromaṇi*.

bhū-gola – a definition of the Sanskrit term offered by A. C. Bhaktivedanta Swami Prabhupāda: "*Bhū* means the earth; *gola* means round."

Bhū-maṇḍala – the earth-disk bisecting the sphere of the Brahmāṇḍa, the Puranic universe. Its central region consists of seven alternating ring-shaped oceans and islands (*dvīpas*), known as Sapta-dvīpa.

Bimalā Prasāda – childhood name of Bhaktisiddhānta Sarasvatī Ṭhākura.

Brahe, Tycho (1546–1601) – Danish astronomer known for his accurate naked-eye observations and planetary measurements. He proposed that the sun revolved around the earth, while the other planets revolved about the sun.

Brahmā – the creator in the Hindu trinity of Brahmā, Viṣṇu, and Śiva.

Brahmagupta (b. 628 CE) – author of the *Brāhma-sphuṭa-siddhānta*, a Sanskrit mathematical astronomical text dated 550 CE. Pṛthūdaka Svāmin commented extensively upon it.

brāhmaṇa – a member of the most intelligent class of men, according to the four *varṇāśrama* Vedic occupational divisions of society.

Brahmāṇḍa – literally Brahmā egg, one of the innumerable universes, described in the *Purāṇas*. It consists of a spherical shell bisected by an earth-disk, Bhū-maṇḍala, of the same radius, with a region of space above and a watery region below.

Caitanya-caritāmṛta – a primary biography describing the philosophy of Caitanya Mahāprabhu, written circa 1557 by Kṛṣṇadāsa Kavirāja.

Caitanya Mahāprabhu – a 15th-century saint from Bengal considered the combined avatar of Rādhā and Kṛṣṇa, who acted in the guise of a devotee. He taught the philosophy of *acintya-bhedābheda-tattva* while propagating pure devotion in the *bhakti* tradition, primarily through *saṅkīrtana* featuring the congregational chanting of the Hare Kṛṣṇa *mantra*.

celestial sphere – an imaginary sphere much larger in diameter than the earth and centered on the earth's center. The stars and planets can be thought of as points projected onto the surface of this sphere. Positions on the celestial sphere are measured by celestial longitude, called right ascension, and celestial latitude, called declination.

Copernicus, Nicolaus (1473–1543) – the Polish astronomer, mathematician, and Catholic canon, famous for establishing a heliocentric theory of planetary motion.

dharma – the Sanskrit term *dharma* has been variously translated as duty, virtue, morality, righteousness or religion, but there is no single English equivalent.

demigods – finite living beings empowered by the Supreme Lord with vast intelligence and influence for the administration of the material universe.

deva – a demigod or godly person.

Dhruvaloka – the spiritual abode of Dhruva Mahārāja, which is said to be situated at the North Celestial Pole.

dvīpa – island.

ecliptic – the geocentric orbit of the sun projected onto the celestial sphere as a great circle. This great circle is tilted at an angle of about 23.5° to the celestial equator. It intersects the celestial equator at two points, known as the vernal and autumnal equinoxes

equinox – one of the two points on the celestial sphere where the ecliptic intersects the equator, known as vernal and autumnal equinoxes, when the length of the day equals the length of the night.

Galileo Galilei (1564–1642) – Italian astronomer, physicist, and engineer who promoted the heliocentric model. He is the first to utilize a telescope for astronomical observations.

Ganges – one of the world's largest rivers, considered sacred within the culture of the Indian subcontinent. Originating in the Himalayas, it flows through India and Bangladesh.

Gaṇeśa – the first son of Śiva and Pārvatī. He is portrayed with the head of an elephant and removes obstacles for those who worship him.

Gaudiya Vaishnava – a follower of Caitanya Mahāprabhu.

gola – spherics, a sphere.

great circle – any circle on the surface of a sphere that is centered on the sphere's center. All other circles on a sphere are smaller than a great circle.

guṇas – the three "modes," or qualities, of the material world: goodness, passion and ignorance.

hasta – a traditional Indian unit of length similar to a cubit, approximately 18 inches, or 45 centimeters.

ISKCON – acronym for the International Society for Krishna Consciousness.

Jai Singh II (1686–1743) – ruler of the Kingdom of Amber, founder of the city of Jaipur. Royal sponsor of mathematical and astronomical research, publications, and translations. He built five Jantar Mantar observatories in India, featuring large masonry instruments designed for naked-eye observation.

Jambūdvīpa – central island (*dvīpa*) of Bhū-maṇḍala having a central mountain (Meru) and nine subdivisions (*varṣas*).

Jñānarāja – author in 1503 CE of the mathematical astronomical text, *Siddhāntasundara*. It is considered the first major *Siddhānta* written after Bhaksarācārya's *Siddhānta-śiromaṇi*.

Jyotir Veda – one of the six Vedāṅgas utilizing applied mathematics and astronomical observation for calculating time. This tradition plays an important role in agriculture, religious ceremony, building construction, and navigation. It is also identified with astrology.

jyotiṣa – astronomy and astrology. One of the six *Vedāṅgas*.

Jyotiḥśāstra – literature identified with the *Jyotir Vedas*. *Sūrya-siddhānta* is a text of this type.

kalpa – the length of a day of Brahma, which lasts 4,320,000,000 years.

karma – material activities, or fruitive work, which incurs reaction.

krośa – an Indian unit of length, one quarter the length of a *yojana*.

Kṛṣṇa – described in the *Bhagavad-gītā* and *Bhāgavata Purāṇa* as the transcendental Supreme Lord.

kṣatriya – a member of the warrior and administrative class in the Vedic social system.

Lokāloka Mountain – circular mountain dividing the inner, illuminated region of Bhū-maṇḍala from the dark, uninhabited, outer region, called Aloka-varṣa, which extends from Lokāloka to the shell of the Brahmāṇḍa.

Mahābhārata – an epic of some 100,000 verses, said to be an encyclopedia of traditional Indian lore.

Mānasottara Mountain – circular mountain in the center of Puṣkaradvīpa which is traversed by the wheel of the sun's chariot.

maṅgalācāraṇa – verses an author composes as an invocation with which to begin a book. The three aims of such an invocation are to offer respects to the author's worshipable deity, to offer blessings to the readers, and to set forth the topic of the book.

mantra – a transcendental sound or Vedic hymn that when repeated delivers consciousness from material illusion.

maṭha – a religious institution.

meridian – any great circle of the earth passing through the poles.

Meru, Mount – the central mountain of Jambūdvīpa. It is shaped like an inverted cone, with the city of demigods, Brahmapurī, on top; and it represents the cosmic axis. Also called Sumeru.

muni – a sage.

nakśāstras – 27 or 28 star constellations, situated roughly along the ecliptic, that play a role in Indian astronomy similar to that of the 12 constellations of the zodiac in the West.

Nārāyaṇa – the four-armed form of Kṛṣṇa who presides over the Vaikuṇṭha planets; Viṣṇu.

paṇḍita – a learned scholar.

Parīkṣit – the son of Abhimanyu who inherited the Kuru throne from Yudhiṣṭhira. Śukadeva Gosvāmī spoke the *Bhāgavatam* to Parīkṣit before an audience of sages.

planisphere – a 2-dimensional map representing the features on a sphere, produced by a stereographic projection.

polestar – a star located near the celestial North Pole. The center of daily rotation of the heavens, as seen from the Northern Hemisphere of the earth.

pramāṇa – evidence or proof.

pratyakṣa – perceivable evidence

Pṛthūdaka Svāmīn – author of a lengthy commentary in 864 CE, on Brahmagupta's *Brāhma-sphuṭa-siddhānta.*

Ptolemy, Claudius – second-century Alexandrian astronomer known for his geocentric epicyclic models for the motion of the planets. These were widely utilized until heliocentric theory displaced it in the 17th century.

Purāṇas – Sanskrit texts from India dealing with ten topics, including the creation and annihilation of the universe, the history of humans and demigods, cosmology, and the relation between the soul and God. The word "*purāṇa*" designates ancient lore.

Rāhu – in Puranic cosmology, a shadow entity or dark planet connected with eclipses of the sun and the moon.

Rāmacandra – an avatar of Kṛṣṇa, or Viṣṇu, as the perfect righteous king.

Rāmānujācārya – the founding *ācārya* of one of the four recognized Vaishnava *sampradāyas.*

sādhu – a religious saint.

sampradāya – a disciplic succession of spiritual preceptors through which Vedic teachings are preserved and transmitted from generation to generation.

sannyāsa – the renounced order, dedicated to cultivating and promoting spiritual life.

Sapta-dvīpa – the central region of Bhū-maṇḍala consisting of seven (*sapta*) alternating ring-shaped oceans and islands (*dvīpa*).

śāstra – revealed scripture; Vedic literature.

Siddhānta – a philosophical text or conclusion, with reference to Vedic literature.

Siddhānta-śiromaṇi – a *Jyotiḥśāstra* written in 1150 CE by the Indian mathematician and astronomer, Bhāskarācārya.

Śiva – destroyer of the cosmos within the Hindu trinity of Brahmā, Viṣṇu, and Śiva.

śloka – a verse; identified with religious texts.

solstice – the time (twice each year) at which the sun reaches its maximum or minimum declination, marked by the longest and shortest days, occurring on about June 21 and December 22.

Śrīdhara Svāmi (1378–1414) – author of the oldest existing commentary on *Śrīmad-Bhāgavatam.*

Śrīmad-Bhāgavatam – same as *Bhāgavata Purāṇa.*

śruti – the *Vedas.*

Stereographic polar projection – a method of mapping the surface of a globe onto a plane, such that any great circle on the sphere is mapped onto a circle in the plane.

Śukadeva Goswāmī – a great renounced sage, son of Dvaipāyana Vyāsa. He heard *Śrīmad-Bhāgavatam* from his father and spoke it to Mahāraja Parīkṣit in an assembly that included Sūta Gosvāmī, who later repeated it to the sages at Naimiṣāraṇya.

Sumeru – see Meru, Mount.

Sūrya – deity associated with the sun, credited with the authorship of the *Sūrya-siddhānta.*

Sūryadāsa (b. 1507 CE) – son of Jñānarāja. Wrote a commentary on the *Bīja-gaṇita* of Bhāskarācārya in 1538 CE, along with other mathematical and astronomical works. He is also known for his poetic and philosophical works.

Sūrya-siddhānta – a Sanskrit astronomical text (*Jyotiḥśāstra*) presenting an epicyclic theory of planetary motion. It also discusses the Indian system of world chronology known as the *yuga* system;

the periodic motions of the sun, moon, and planets; distances from the earth to the sun, the moon, and the planets; rules for calculating the times of solar and lunar eclipses; and cosmology and geography.

Temple of the Vedic Planetarium (TOVP) – ISKCON's main temple located in Mayapur, West Bengal. Construction began in 2010, and when complete will be the world's largest Hindu temple.

Vaishnava – a devotee of the Supreme Lord.

Vaṁśīdhara – author of the *Bhāgavatam* commentary, *Bhavārtha-dīpikā-prakāśa.*

varṇas – the four social and occupational divisions of Vedic society.

vārsa – any one of the nine subdivisions of Jambūdvīpa, separated from one another by mountain ranges.

Vedāṅga – one of the six recognized supporting disciplines of the *Vedas.*

Vedas – ancient Sanskrit texts of revealed knowledge which are seen as authoritative within Hindu culture.

Varāhamihira (c. 505–c. 587) – author of the *Pañcasiddhāntikā*, describing information from older astronomical texts, as well as the encyclopedic *Bṛhat-saṁhitā*, covering a wide range of topics drawn from Indian culture during this period.

virāṭ-rūpa – the universal form of the Supreme Lord exhibited within the material cosmos.

Viṣṇu – described in the *Bhāgavata Purāṇa* as the transcendental Supreme Lord, an expansion of Kṛṣṇa. He is also known as the maintainer in the Hindu trinity of Brahmā, Viṣṇu, and Śiva.

Viṣṇu Purāṇa – one of the eighteen major *Purāṇas*. It contains an extensive section on cosmology which complements the cosmological section of the *Bhāgavatam.*

Viśvanātha Cakravartī – a prominent Gaudiya Vaishnava *ācārya* in the line of Śrīla Narottama dāsa Ṭhākura. He wrote commentaries in the mid-17th century on *Śrīmad-Bhāgavatam, Bhagavad-gītā,* as well as on the works by followers of Caitanya Mahāprabhu.

Vṛndāvana – the transcendental abode of Lord Kṛṣṇa. It is also called Goloka Vṛndāvana or Kṛṣṇaloka. The town of Vrindavan in the Mathura District of Uttar Pradesh, India, is considered a manifestation on earth of Kṛṣṇa's abode in the spiritual world.

Vyāsadeva – the compiler of the *Vedas* and author of the *Purāṇas, Mahābhārata*, and *Vedānta-sūtra*.

yoga – spiritual discipline meant to link its practitioner with the Supreme.

yogi – a practitioner of a yoga system.

yojana – an Indian unit of length used to express distances in the *Bhāgavatam*. Different standard *yojana* lengths have been described, including one of about eight miles.

Made in United States
Troutdale, OR
11/03/2026

56847410R00159